Mining Latent Entity Structures

Synthesis Lectures on Data Mining and Knowledge Discovery

Editor
Jiawei Han, *University of Illinois at Urbana-Champaign*
Lise Getoor, *University of Maryland*
Wei Wang, *University of North Carolina, Chapel Hill*
Johannes Gehrke, *Cornell University*
Robert Grossman, *University of Chicago*

Synthesis Lectures on Data Mining and Knowledge Discovery is edited by Jiawei Han, Lise Getoor, Wei Wang, Johannes Gehrke, and Robert Grossman. The series publishes 50- to 150-page publications on topics pertaining to data mining, web mining, text mining, and knowledge discovery, including tutorials and case studies. Potential topics include: data mining algorithms, innovative data mining applications, data mining systems, mining text, web and semi-structured data, high performance and parallel/distributed data mining, data mining standards, data mining and knowledge discovery framework and process, data mining foundations, mining data streams and sensor data, mining multi-media data, mining social networks and graph data, mining spatial and temporal data, pre-processing and post-processing in data mining, robust and scalable statistical methods, security, privacy, and adversarial data mining, visual data mining, visual analytics, and data visualization.

Graph Mining: Laws, Tools, and Case Studies
D. Chakrabarti and C. Faloutsos
2012

Mining Heterogeneous Information Networks: Principles and Methodologies
Yizhou Sun and Jiawei Han
2012

Privacy in Social Networks
Elena Zheleva, Evimaria Terzi, and Lise Getoor
2012

Community Detection and Mining in Social Media
Lei Tang and Huan Liu
2010

Ensemble Methods in Data Mining: Improving Accuracy Through Combining Predictions
Giovanni Seni and John F. Elder
2010

Modeling and Data Mining in Blogosphere
Nitin Agarwal and Huan Liu
2009

Mining Latent Entity Structures

Chi Wang and Jiawei Han

ISBN: 978-3-031-00779-8   paperback
ISBN: 978-3-031-01907-4   ebook

DOI 10.1007/978-3-031-01907-4

A Publication in the Springer series
SYNTHESIS LECTURES ON DATA MINING AND KNOWLEDGE DISCOVERY

Lecture #10
Series Editors: Jiawei Han, *University of Illinois at Urbana-Champaign*
            Lise Getoor, *University of Maryland*
            Wei Wang, *University of North Carolina, Chapel Hill*
            Johannes Gehrke, *Cornell University*
            Robert Grossman, *University of Chicago*
Series ISSN
Print 2151-0067   Electronic 2151-0075

Mining Latent Entity Structures

Chi Wang
Microsoft Research

Jiawei Han
University of Illinois at Urbana-Champaign

SYNTHESIS LECTURES ON DATA MINING AND KNOWLEDGE DISCOVERY #10

ABSTRACT

The "big data" era is characterized by an explosion of information in the form of digital data collections, ranging from scientific knowledge, to social media, news, and everyone's daily life. Examples of such collections include scientific publications, enterprise logs, news articles, social media, and general web pages. Valuable knowledge about multi-typed entities is often hidden in the unstructured or loosely structured, interconnected data. Mining latent structures around entities uncovers hidden knowledge such as implicit topics, phrases, entity roles and relationships.

In this monograph, we investigate the principles and methodologies of mining latent entity structures from massive unstructured and interconnected data. We propose a text-rich information network model for modeling data in many different domains. This leads to a series of new principles and powerful methodologies for mining latent structures, including (1) latent topical hierarchy, (2) quality topical phrases, (3) entity roles in hierarchical topical communities, and (4) entity relations. This book also introduces applications enabled by the mined structures and points out some promising research directions.

KEYWORDS

information networks, text mining, link analysis, topic modeling, phrase extraction, role discovery, clustering, ranking, relationship mining, probabilistic models, real-world applications, efficient and scalable algorithms

Contents

Acknowledgments

The authors would like to acknowledge Marina Danilevsky, Ahmed El-kishky, and Rui Li for their tremendous research collaboration.

Wang's work was supported in part by a Microsoft Research Ph.D. fellowship.

Han's work was supported in part by the U.S. Army Research Lab. under Cooperative Agreement No. W911NF-09-2-0053 (NSCTA), the Army Research Office under Cooperative Agreement No. W911NF-13-1-0193, National Science Foundation IIS-1017362, IIS-1320617, and IIS-1354329, HDTRA1-10-1-0120, and trans-NIH Big Data to Knowledge (BD2K) initiative grant 1U54GM114838 awarded by NIGMS, and MIAS, a DHS-IDS Center for Multimodal Information Access and Synthesis at UIUC.

Chi Wang and Jiawei Han
March 2015

CHAPTER 1

Introduction

1.1 MOTIVATION

The success of database technology is largely attributed to the efficient and effective handling of *structured data*. The construction of a well-structured database is often the premise of subsequent applications. However, the explosion of "big data" poses great challenges on this practice since the real world data are largely unstructured, or loosely structured. It is crucial to uncover latent structures of real-world entities, such as the topics and communities they are involved in, the roles the entities play in these topics and communities, and the relations they potentially have with each other. By mining massive unstructured or loosely structured data associated with entities, one can construct semantically rich structures which reveal the relationships among entities. The uncovered structures facilitate browsing information and retrieving knowledge from the data.

Mining latent entity structures will enhance knowledge engineering effectively in many applications. For example, mining entity structures hidden in billions of web pages will turn extensive web data to knowledge that will enrich open-domain knowledge-bases. Mining entity structures hidden in social media will help reorganize scattered information from hundreds of millions of individuals and improve the social network services. In news data, the topics, as well as the entity relations, are buried in the text rather than in the form of relational tuples. Mining such news data will enable us to extract multiple types of entities like people, locations, organizations, and events, for effective news understanding and analysis. In a bibliographic database like DBLP[1] or PubMed,[2] research papers are explicitly linked with authors, venues, and terms. Many interesting semantic relationships, such as advisor-advisee between authors, are hidden in the publication records; moreover, the research topics of authors, venues and terms are also hidden or unorganized, preventing insightful organization of the entities. Mining hidden research network structures will help scientific research tremendously.

[1]http://www.informatik.uni-trier.de/
[2]http://www.ncbi.nlm.nih.gov/pubmed/

1.2   DATA MODEL: A TEXT-RICH HETEROGENEOUS INFORMATION NETWORK MODEL

Many real datasets collected in our world contain both unstructured data and some (weak) structures, as shown in the following examples.

Example 1.1   Research publications.   A bibliographic database like DBLP and PubMed contains unstructured paper text (titles, abstracts, full text, etc.), as well as structures, in the sense that every paper is explicitly linked with authors, venues, and years.

Example 1.2   News articles.   A collection of news articles contains unstructured news text (titles, snippets, full text, etc.), as well as weak structures, in the sense that every news article is implicitly linked with named entities like persons, locations, and organizations.

Example 1.3   Social media.   A social media website like Twitter contains unstructured tweets, as well as structures, in the sense that every tweet is linked with tweeters, URLs or hashtags. There are also links among the tweeters by their following relationship. In addition, every tweeter is linked to his/her profile. The profile may have structured fields such as interests and locations, while each field may contain unstructured text.

To systematically model unstructured and interconnected data, we model our datasets using a text-rich heterogeneous information network model, defined as follows.

Definition 1.4   Text-rich heterogeneous information network.   A text-rich heterogeneous information network can be formally represented by a tuple $H = (\{\mathcal{V}_t\}, \{\mathcal{E}_{x,y}\}, \mathcal{D})$, where $\mathcal{V}_t$ is the set of type-t nodes, $\mathcal{E}_{x,y}$ is the set of link weights, linking between the nodes of types x and y (x and y may be identical), and $\mathcal{D} = \{d_v\}$ is the set of documents. Note that d_v is the document attached to node v, and d_v can be empty if node v is not attached with any text.

The computer science research publication dataset, DBLP, for example, can be formally represented as follows.

Example 1.5   The DBLP network.   The network has three types of nodes: $\mathcal{V}_1$ for papers, $\mathcal{V}_2$ for authors, and $\mathcal{V}_3$ for venues, and two types of links: $\mathcal{E}_{1,2}$ for the links between papers and authors; and $\mathcal{E}_{1,3}$ for the links between paper and its publication venue. A short document d_i, which is the title of the paper, is attached to every paper node $v_i^1 \in \mathcal{V}_1$. Note that only the titles are available as unstructured text in the DBLP database.

Our data model is a generalization of several simpler forms of data: (i) text-only corpus, in which $\mathcal{E} = \emptyset$ and $|\mathcal{D}| = |\mathcal{V}|$; (ii) text-absent heterogeneous network, in which $\mathcal{D} = \emptyset$; and (iii) textual corpus with meta data or star-schema network, in which all the links share a central type of nodes.

The text-rich heterogeneous information network will be the input from which we mine latent entity structures. The algorithms studied in this book do not always require the existence of both text and links, so they can work when certain information is missing.

1.3   LATENT ENTITY STRUCTURE

To understand what structures are important, we analyze examples of knowledge that people are interested in discovering from the text-rich information networks.

Example 1.6   From research publications, people often ask the following questions:

- Who are the leading researchers in Computer Science?

- (continued) What are their specializations?

- Who are the peer researchers of Jure Leskovec?

- With whom will he collaborate?

To answer questions like these, it is beneficial to discover latent research topics in multiple granularity and related authors, as well as discovering relations among researchers.

Example 1.7   From news articles, people ask:

- Who are the top 10 active politicians regarding the gun control issue?;

- What are important events related to the U.S.-Cuba relationship?;

and so on. In order to answer these questions, one must understand the latent topics in news, and the related entities like people and events.

Example 1.8   From social networks, people ask:

- Is Angelina Jolie a relative, schoolmate, colleague, or other friend of Matthew Crocker?;

- What is Andrew's interest and expertise?;

- Who are the most influential Twitter users on the topic of the movie "The Interview?";

and so on. These questions require one to understand the latent topics in social media, and the relations between users in the social network.

In general, the latent topics, the roles of entities in the topics, and relations between entities are the main latent structures to be covered in this book. With these structures, a number of applications that rely on structured database queries or information network analysis can be enabled. We discuss a few applications in Chapter 7.

Formally, the latent structures can be defined as tuples in the form of (R, A, B, x), where A and B represent topics, phrases, or entities, R represent a relation between A and B, and x is a numeric value to quantify the relation. The following cases are covered in this book.

- A is a topic, B is a topic, R represents that the relation A is a subtopic of B, and $x \in [0, 1]$ represents the probability of seeing topic A given topic B.

- A is a phrase, B is a topic, R represents that the relation A is a topical phrase in B, and $x \in [0, 1]$ represents the representativeness of phrase A in topic B.

- A is an entity, B is a topic, R represents that A plays a certain role in B, and $x \in [0, 1]$ implies the importance of entity A in that role.

- A is an entity, B is an entity, R represents the a certain relationship between the two entities, and $x \in [0, 1]$ implies the likelihood of that relationship.

1.4   THE MINING FRAMEWORK

This section presents a framework for mining hierarchical topics and uncovering hidden entity relations. It comprises the following functional modules: *hierarchical topic and community discovery*, *topical phrase mining*, *entity topical role analysis*, and *entity relationship mining*.

1.4.1   HIERARCHICAL TOPIC AND COMMUNITY DISCOVERY

As shown in the example questions, knowledge is often confined to a certain scope in a large network. For example, to mine opinion leaders, we need to specify a scope since different communities or sub-communities may have different opinion leaders. By mining topical communities from a network, one can use a community as its contextual scope. For example, the communities formed by research topics can provide suitable context for analyzing important contributors in each community.

In particular, we explore the hierarchical structure of a topic. Hierarchy is an effective and efficient structure for both humans and machines to use in many applications. This module is to construct a topical hierarchy from a text-rich heterogeneous information network.

1.4.2   TOPICAL PHRASE MINING

Single words often contain lots of ambiguity in comparison with phrases which are often less ambiguous for humans to comprehend. For example, the topic of query processing may be described by the phrases {'query processing', 'query optimization', 'top-k queries', ...}, while its parent topic can be "databases," described by {'query processing', 'database systems', 'concurrency control', ...}. This module mines phrases from the text automatically and ranks them according to their representativeness in each topic.

1.4.3   ENTITY TOPICAL ROLE ANALYSIS

After a topical community is discovered, one can analyze the roles of entities in a desired scope. There are two types of questions to be answered.

- Type-A: For a given topical community, *what is the role of a specific entity* in the community? For instance, which topics within the community get published in a particular conference? Or, which specific topics within the community does an author contribute to?

- Type-B: Given a topical community and an entity type, *which entities play the most important roles* in the community? For example, an author's contribution to the topics of a community (by publishing papers) represents the author's role in that community.

1.4.4   ENTITY RELATIONSHIP MINING

This module is to mine latent relationships among entities from multiple signals, such as text and links.

In summary, the main features of the mining framework are as follows.

- *Leveraging data redundancy*: The massive unstructured data contain repetitive patterns. Our text mining techniques leverage these patterns, instead of specific linguistic analysis for individual documents or sentences. Therefore, it is generally applicable to a large amount of data, and its power gains when the size of data increases.

- *Recursive hierarchy construction*: The hierarchy is constructed in a top-down order. One can recursively apply our method to expand the hierarchy. It is flexible to revise part of the hierarchy with other remaining parts intact.

- *Phrase-represented topic*: Each topic is represented by a ranked list of topical phrases, such that a child topic is a subset of its parent topic. The ranked list of phrases is much more interpretable than the ranked list of unigrams as the traditional visualization.

- *Entity-embedded topic*: The linked entity information enhances the topic discovery solely based on textual information. For example, the authors and venues linked to each paper help finding its topics. In our framework, each topic can be enriched with ranked lists of entities, which has two advantages: (i) the entities provide additional informative context for each topic in the hierarchy; and (ii) the entity positions in the hierarchy are discovered and ready to query as we construct the hierarchy.

- *Dependency modeling of entity relations*: Utilizing heterogeneous links and text information, our method is able to capture a variety of semantic signals, including constraints and dependencies, to recover a certain relationship with the best known accuracy.

The rest of the book is organized as follows. The above four modules are presented in Chapters 2–5. Chapter 6 focuses on solving a computational challenge to scale up the method. Finally, Chapter 7 discusses the applications and the research frontier.

CHAPTER 2

Hierarchical Topic and Community Discovery

Automated organization of topics from unstructured and interconnected data at multiple levels of granularity is an important problem in knowledge engineering with many valuable applications, such as information summarization, search, and online analytical processing (OLAP). A student could familiarize herself with a new domain by perusing such a hierarchy and quickly learning the topics of the domain. Or, a researcher could discover which phrases are representative of his topic of interest, assisting his search for relevant work and discovering potential subtopics to focus on.

With a vast amount of data and dynamic changes in users' need, it is too costly to rely on human experts to do manual annotation and provide ready-to-use topical hierarchies. Thus, it is critical to create a robust method for *automated construction of high-quality topical hierarchies from text-rich heterogeneous information networks*, a key task in the framework of mining latent entity structures.

This chapter proposes a method that discovers hierarchical topics recursively, that is, grows the topical tree gradually from the root. A key operation of the methodology is to mine subtopics of a topic represented by a leaf vertex in the current tree, so that vertices that represent subtopics will be added as children of this vertex. Repeating this operation will grow the tree recursively in a top-down manner. The tree constructed at any time during this process encodes the discovered topics so far, and can be used as output of the mined hierarchical topics.

The recursive strategy has the following advantages.

- The parent-child relation is ensured during the recursion. In non-recursive methods, the parent-child relation may not be obvious due to sampling along long paths.

- One can stop expanding the hierarchy after a number of growing operations as needed. This contrasts with non-recursive methods where all of the topics must be inferred before the algorithm stops.

- It is flexible to revise part of a hierarchy with the other remaining parts intact. In non-recursive methods, the change of the shape of a hierarchy may result in an entirely different hierarchy.

- As a network-based analysis method, it can be extended naturally to heterogeneous networks, with or without text. The non-recursive methods have intricate generative processes and are hard to be extended to other networks.

For the data where no text information is available, the method can be applied to find hierarchical community structures only based on links.

2.1 GENERATIVE MODEL FOR TEXT OR HOMOGENEOUS NETWORKS

In this section we propose a method for recursively discovering hierarchical topics from text only. The method is based on analysis with term co-occurrence networks, which are converted from the collection of documents. The analysis technique can be applied to generic homogeneous networks (with only one type of nodes and links). For convenience we index all the unique words in this corpus using a vocabulary of V words.

Given a text corpus, our goal is to construct a topical hierarchy. A topical hierarchy is a tree of topics, where each child topic is about a more specific theme within the parent topic. Statistically, a topic t is characterized by a probability distribution over words ϕ_t. $\phi_{t,x} = p(x|t) \in [0, 1]$ is the probability of seeing the word x in topic t, and $\sum_{x=1}^{V} \phi_{t,x} = 1$. For example, in a topic about the database research area, the probability of seeing "database," "system," and "query" is high, and the probability of seeing "speech," "handwriting," and "animation" is low. This characterization is advantageous in statistical modeling of text, but is weak in human interpretability, due to two reasons. First, unigrams (i.e., single words) may be ambiguous, especially across specific topics. Second, the probability $p(x|t)$ reflects the popularity of a word x in the topic t, but not its discriminating power. For example, a general word "algorithm" may have higher probability than a discriminative word "locking" in the database topic.

For these reasons, we choose to enhance the topic representation with ranked phrases. Phrases reduce the ambiguity of unigrams. And the ranking should reflect both their popularity and discriminating power for a topic. For example, the database topic can be described as: {"database systems," "query processing," "concurrency control," ...}. A phrase can appear in multiple topics, though it will have various ranks in them.

Formally, we define a topical hierarchy as follows.

Definition 2.1 Topical hierarchy. A topical hierarchy is defined as a tree $\mathcal{T}$ in which each node is a topic. Every non-leaf topic t has C_t child topics. Topic t is characterized by a probability distribution over words ϕ_t, and visualized as an ordered list of phrases $\mathcal{P}_t = \{P_{t,1}, P_{t,2}, \ldots\}$, where $P_{t,i}$ is the phrase ranked at the i-th position for topic t.

The number of child topics C_t of each topic can be specified as input, or decided automatically by the construction method. In both cases, we assume it is bounded by a small number K, such as 10. This is for efficient browsing of the topics. The number K is named the *width* of the tree $\mathcal{T}$.

For convenience, we denote a topic using the top-down path from root to this topic. The root topic is denoted as o. Every non-root topic t is denoted by $\pi_t \odot \chi_t$, where π_t is the notation of its parent topic, and χ_t is the index of t among its siblings. For example, the topic $t1$ in Figure 2.1 is denoted as $o \odot 1$, and its child $t5$ is denoted as $o \odot 1 \odot 2$. The *level* h_t of a topic t is defined to be the number of "$\odot$" in its notation. So the root topic is at level 0, and $t5$ is at level 2. The *height* h of a tree is defined to be the maximal level over all the topics in the tree. Clearly, the total number T of topics is upper bounded by $\frac{K^{h+1}-1}{K-1}$.

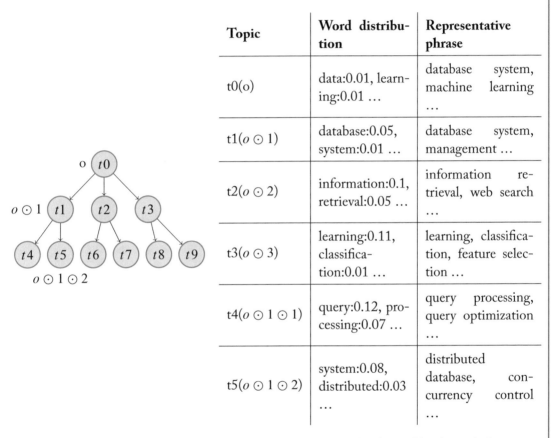

Topic	Word distribution	Representative phrase
t0(o)	data:0.01, learning:0.01 …	database system, machine learning …
t1($o \odot 1$)	database:0.05, system:0.01 …	database system, management …
t2($o \odot 2$)	information:0.1, retrieval:0.05 …	information retrieval, web search …
t3($o \odot 3$)	learning:0.11, classification:0.01 …	learning, classification, feature selection …
t4($o \odot 1 \odot 1$)	query:0.12, processing:0.07 …	query processing, query optimization …
t5($o \odot 1 \odot 2$)	system:0.08, distributed:0.03 …	distributed database, concurrency control …

Figure 2.1: An example of the topical hierarchy. Each topic can be denoted by the path from root topic to it.

Now we present a network analysis technique for topical hierarchy construction. Every topic node t in the topical hierarchy is associated with a term co-occurrence network G^t. The root node o is associated with the term co-occurrence network G^o constructed from the collection of documents. G^o consists of V nodes and a set of links $\mathcal{E}$. A node $v \in [V]$ represents a term, and a link (i, j) between two nodes represents a co-occurrence of the two terms in a document. The number of links $e_{ij} \in \mathcal{E}$ between two nodes i and j is equal to the number of co-occurrences of the two terms. For every non-root node $t \neq o$, we construct a subnetwork G^t by clustering the term co-occurrence network of its parent π_t. G^t has all of the nodes from G^{π_t}, but only those links belonging to the particular subtopic t.

We choose to use term co-occurrence network for topic analysis instead of document-term topic modeling because it naturally supports recursive mining: the clustering result for one topic can be used as the input when further partitioning the topic into subtopics. We name this method CATHY (Construct A Topical HierarchY). It produces a topical hierarchy in a top-down, recursive manner.

Step 1. Construct the term co-occurrence network G^o from the document collection. Set $t = o$.

Step 2. For a topic t, cluster the term co-occurrence network G^t into subtopic subnetworks $G^{t/z}, z \in [C_t]$.

Step 3. Recursively apply Step 2 to each subtopic $t \odot z, z \in [C_t]$ to construct the hierarchy in a top-down fashion.

In the following, we introduce the process of clustering for one topic t. We assume $C_t = k$. The value of k can be either specified by users or chosen using a model selection criterion such as the Bayesian Information Criterion [Schwarz, 1978].

In the term "co-occurrence" network G^t, we assume every co-occurrence of two terms i and j is attributed to a topic $t \odot z, z \in [C_t] = [k]$. Thus, the total link frequency e_{ij} between i and j is the summation of the number of links between i and j in each of the k topics: $e_{ij}^t = \sum_{z=1}^{k} e_{ij}^{t \odot z}$. The goal is thus to estimate $e_{ij}^{t \odot z}$ for $z = 1 \ldots k$. This is different from other network analysis approaches which hard partition their nodes or edges.

To generate a topic-$t \odot z$ link, we first generate one end node i following a multinomial distribution $p(i|t \odot z) = \phi_{t \odot z, i}$, and then generate the other end node j with the same multinomial distribution $p(j|t \odot z) = \phi_{t \odot z, j}$. The probability of generating a topic-$t \odot z$ link (i, j) is therefore $p(i|t \odot z)p(j|t \odot z) = \phi_{t \odot z, i}\phi_{t \odot z, j}$.

With this generative assumption for each individual link, we can derive the distribution of topical frequency for any two terms (i, j). If we repeat the generation of topic-$t \odot z$ links for $\rho_{t \odot z}$ iterations, then the chance of generating a particular topic-$t \odot z$ link between i and j can be modeled as a Bernoulli trial with the success probability $\phi_{t \odot z, i}\phi_{t \odot z, j}$. When $\rho_{t \odot z}$ is large, the total number of successes $e_{ij}^{t \odot z}$ approximately follows a Poisson distribution $Pois(\rho_{t \odot z}\phi_{t \odot z, i}\phi_{t \odot z, j})$.

Now we can write down the generative model for random variables $e_{ij}^{t \odot z}$ with parameters $\rho_{t \odot z}, \phi_{t \odot z}$:

$$e_{ij}^{t \odot z} \sim Poisson(\rho_{t \odot z} \phi_{t \odot z,i} \phi_{t \odot z,j}), z = 1, \ldots, k \tag{2.1}$$

$$\sum_{i=1}^{V} \phi_{t \odot z,i} = 1, \quad \phi_{t \odot z,i} \geq 0, \rho_{t \odot z} \geq 0. \tag{2.2}$$

The constraints guarantee a probabilistic interpretation. According to the *expectation* property of the Poisson distribution, $E(e_{ij}^{t \odot z}) = \rho_{t \odot z} \phi_{t \odot z,i} \phi_{t \odot z,j}$. Also, according to the *additive* property of expectations,

$$E(\sum_{i,j} e_{ij}^{t \odot z}) = \sum_{i,j} \rho_{t \odot z} \phi_{t \odot z,i} \phi_{t \odot z,j} = \rho_{t \odot z} \sum_{i} \phi_{t \odot z,i} \sum_{j} \phi_{t \odot z,j} = \rho_{t \odot z}.$$

In other words, $\rho_{t \odot z}$ is the total expected number of links in topic $t \odot z$.

One important implication due to the *additive* property of Poisson distribution is that

$$e_{ij}^{t} = \sum_{z=1}^{k} e_{ij}^{t \odot z} \sim Poisson(\sum_{z=1}^{k} \rho_{t \odot z} \phi_{t \odot z,i} \phi_{t \odot z,j}). \tag{2.3}$$

So given the model parameters, the probabilities of all observed links are

$$p(\{e_{ij}^{t}\}|\phi, \rho) = \prod_{i,j \in [V]} p(e_{ij}^{t}|\phi_i, \phi_j, \rho)$$

$$= \prod_{i,j \in [V]} \frac{(\sum_{z=1}^{k} \rho_{t \odot z} \phi_{t \odot z,i} \phi_{t \odot z,j})^{e_{ij}^{t}} \exp(-\sum_{z=1}^{k} \rho_{t \odot z} \phi_{t \odot z,i} \phi_{t \odot z,j})}{e_{ij}^{t}!}. \tag{2.4}$$

In this model, the observed information is the total number of links between every pair of nodes, including zero links and self-links. The parameters which must be learned are the role of each node in each topic $\phi_{t \odot z,i}, i \in [V], z = 1, \ldots, k$, and the expected number of links in each topic $\rho_{t \odot z}$. The total number of free parameters to learn is therefore kV. We learn the parameters by the *Maximum Likelihood* (ML) principle: find the parameter values that maximize the likelihood in Eq. (2.4). We use the Expectation-Maximization (EM) algorithm that iteratively infers the model parameters:

$$E - step : \quad \hat{e}_{ij}^{t \odot z} = e_{ij}^{t} \frac{\rho_{t \odot z} \phi_{t \odot z,i} \phi_{t \odot z,j}}{\sum_{x=1}^{k} \rho_{t \odot x} \phi_{t \odot x,i} \phi_{t \odot x,j}} \tag{2.5}$$

$$M - step :$$

$$\rho_{t \odot z} = \sum_{i,j} \hat{e}_{ij}^{t \odot z} \tag{2.6}$$

$$\phi_{t \odot z,i} = \frac{\sum_{j} \hat{e}_{ij}^{t \odot z}}{\rho_{t \odot z}}. \tag{2.7}$$

Intuitively, the E-step calculates the expected number of links $\hat{e}_{ij}^{t\odot z}$ in each topic $t \odot z$ between the terms i and j: the ratio of $\hat{e}_{ij}^{t\odot z}$ to e_{ij}^{t} is proportional to its Poisson parameter $\rho_{t\odot z}\phi_{t\odot z,i}\phi_{t\odot z,j}$. The M-step calculates the ML parameter estimates: $\phi_{t\odot z,i}$ is the ratio of the total number of links in topic $t \odot z$ where one end node is i and $\rho_{t\odot z}$, which is the sum of the total expected number of links in topic $t \odot z$.

We update $\hat{e}_{ij}^{t\odot z}, \phi_{t\odot z,i}, \rho_{t\odot z}$ at each iteration. Note that if $e_{ij}^{t} \notin \mathcal{E}$, we do not need to calculate $\hat{e}_{ij}^{t\odot z}$ because it equals 0. Therefore, the time complexity for each iteration is $\mathcal{O}((|\mathcal{E}| + V)k) = \mathcal{O}(|\mathcal{E}|k)$. Like other EM algorithms, the solution converges to a local maximum and the result may vary with different initializations. The EM algorithm may run multiple times with random initializations to find the solution with the best likelihood. We empirically find that the EM algorithm generally requires hundreds of iterations to converge. A more scalable method will be introduced in Chapter 6.

It is important to note that this method naturally supports top-down hierarchical clustering. To further discover subtopics of a topic, we can extract the subnetwork where $\mathcal{E}^{t\odot z} = \{\hat{e}_{ij}^{t\odot z}|\hat{e}_{ij}^{t\odot z} \geq 1\}$ (the expected number of links attributed to that topic, ignoring values less than 1) and then apply the same generative model on the subnetwork. This process repeats recursively until the desired hierarchy is constructed.

After the hierarchy is discovered, we can use the technique in next chapter to visualize the topics with topical phrases. A sample output with the DBLP publication titles is visualized in Figure 2.2.

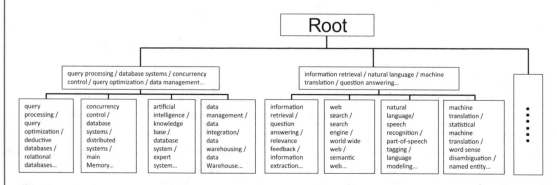

Figure 2.2: Sample output from CATHY—topical hierarchy of text only. Each node in the hierarchy contains a ranked list of phrases.

2.2   GENERATIVE MODEL FOR HETEROGENEOUS NETWORK

As we discussed in Chapter 1, digital documentary data collections often contain additional information beyond plain text. For example, a research paper is linked to its authors and the venue it was published. A tweet is linked to its twitter and the hashtags or URLs mentioned in the tweet. The text linked with multi-typed objects (authors, venues, twitters, hashtags, etc.) form text-rich heterogeneous information networks.

Term	Author	Venue
data	divesh srivastava	SIGMOD/
database	jeffrey f. naughton	ICDE
queries	christos faloutsos	VLDB
system	raghu ramakrishnan	PODS
query...	surajit chaudhuri...	CIKM...

Term	Author	Venue
model	w. bruce croft	SIGIR
retrieval	chengxiang zhai	ACL
learning	james allan	CIKM
information	maarten de rijke	IJCAI
text...	c. lee giles...	AAAI...

• • • • • •

Figure 2.3: Sample output from NetClus—clusters of multi-typed entities. Each rounded rectangle represents one cluster, containing a ranked list of unigrams and two ranked lists of entities.

Few approaches utilize link information from typed entities that may be present in the data. Conversely, existing methods for heterogeneous network analysis and topic modeling have demonstrated that multiple types of linked entities improve the quality of topic discovery (e.g., NetClus [Sun et al., 2009]), but these methods are not designed for finding hierarchical structures (see Figure 2.3 for an example output of NetClus).

In this section, we develop a method that makes use of both textual information and heterogeneous linked entities to automatically construct multi-typed topical hierarchies. It is an extension of the generative model for text in the last section.

First, we extend the definition of text-based topical hierarchy to heterogeneous topical hierarchy.

Definition 2.2   Heterogeneous Topical Hierarchy.   A heterogeneous topical hierarchy is defined as a tree $\mathcal{T}$ in which each node is a topic. Topic t is characterized by a probabilistic distribution for each type of entities, including terms. We assume each node type x has a multinomial distribution $\phi^{x,t\odot z}$ in each subtopic $t \odot z, z \in [k]$, such that $\phi^{x}_{t\odot z,i}$ is the importance of node v^{x}_{i} in topic $t \odot z$, subject to $\sum_{i} \phi^{x}_{t\odot z,i} = 1$. Each node type x also has a multinomial distribution $\phi^{x}_{t\odot 0}$ for the background topic.

Example 2.3   In a computer science publication network, each of the three node types "term," "author," and "venue" has a ranking distribution in each topic in the hierarchy. The distributions for a hypothetical topic about database may be: (i) term - {database: 0.01, system: 0.005, query: 0.004, ...}; (ii) author - {Sujarit Chaudhuri: 0.03, Jeffery F. Naughton: 0.02, ...}; and (iii) venue - {SIGMOD: 0.2, VLDB: 0.25, ...}

Second, we extend the definition of edge-weighted network associated with each topic node from homogeneous network to heterogeneous network. Formally, every topic node t in the topical hierarchy is now associated with an edge-weighted network $G^{t} = (\{\mathcal{V}^{t}_{x}\}, \{\mathcal{E}^{t}_{x,y}\})$, where

$\mathcal{V}_x^t$ is the set of type-x nodes in topic t, and $\mathcal{E}_{x,y}^t$ is the set of link weights between type x and type y nodes (x and y may be identical) in topic t. $e_{i,j}^{x,y,t} \in \mathcal{E}_{x,y}^t$ represents the weight of the link between node v_i^x of type x and node v_j^y of type y.

To construct the network G^o for the root topic o, we convert the text-attached heterogeneous information network $H = (\{\mathcal{V}_x\}, \{\mathcal{E}_{x,y}\}, \mathcal{D})$ into a collapsed network. Assume there are m node types and m^2 link types in H. Type-1 nodes are documents. G^o is obtained by converting the documents into term co-occurrence links and adding them into the network in H. $G^o = (\{\mathcal{V}_x\}_{x=2}^m \cup \{\mathcal{V}_{m+1} = [V]\}, \{\mathcal{E}_{x,y}\}_{x,y=2}^m \cup \{\mathcal{E}_{x,m+1}\}_{x=2}^{m+1})$. The $(m+1)$-th type of nodes in G^o are terms. If an entity is linked to a document in H, then it is linked with all the words in that document. The link weight between an entity and a word in G^o is equal to the summation of the link weight between the entity and all the documents in H.

Example 2.4 We can construct a collapsed network from the research publications, with $m = 3$ types of nodes: term, author, and venue; and $l = 5$ types of links: term-term, term-author, term-venue, author-author, author-venue. The link weight between every two nodes is equal to the number of papers where the two objects co-occur.

When there is only text information, the collapsed network reduces to term co-occurrence network, as we discussed in the last section. When there is no text information, $G^o = H$.

For every non-root node $t \neq o$, we construct a subnetwork G^t by clustering the network G^{π_t} of its parent π_t. G^t inherits the nodes from G^{π_t}, but contains only the fraction of the original link weights that belongs to the particular subtopic t. In other words, we softly partition the link weights in G^t into subtopics. Figure 2.4 visualizes the weight of each link in each network and subnetwork by line thickness (disconnected nodes and links with weight 0 are omitted).

This method is named CATHYHIN (construct a topical hierarchy from heterogeneous information network). It outputs a heterogeneous topical hierarchy in a top-down, recursive way.

Step 1. Construct the edge-weighted network G^o. Set $t = o$.

Step 2. For a topic t, cluster the network G^t into subtopic subnetworks $G^{t \odot z}, z \in [C_t]$ using a generative model.

Step 3. Recursively apply Step 2 to each subtopic $t \odot z, z \in [C_t]$ to construct the hierarchy in a top-down fashion.

After the hierarchy is discovered, we can use the technique in Chapter 3 to visualize the topics with topical phrases, and analyze entity roles using the technique in Chapter 4. A sample output for the DBLP network is visualized in Figure 2.5.

In Section 2.2.1, we describe a unified generative model and present an inference algorithm with a convergence guarantee. In Section 2.2.2, we further extend the approach to allow different link types to have different degrees of importance in the model (allowing the model to, for example, decide to pay unequal attention to term co-occurrence information and co-author information).

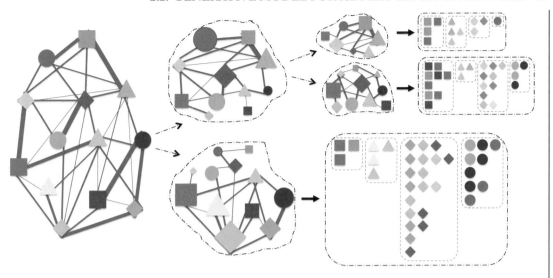

Figure 2.4: An illustration of the CATHYHIN framework. **(L)** Step 1: CATHYHIN analyses a node-typed and edge-weighted network, with no central star objects. **(M)** Step 2: A unified generative model is used to partition the edge weights into clusters and rank single nodes in each cluster (here, node rank within each node type is represented by variations in node size). **(R bottom)** Step 3: Patterns of nodes are ranked within each cluster, grouped by type. This step is discussed in Chapter 3. **(R top)** Step 4: Each cluster is also an edge-weighted network, and is therefore recursively analyzed. The final output is a hierarchy, where the patterns of nodes of each cluster have a ranking within that cluster, grouped by type.

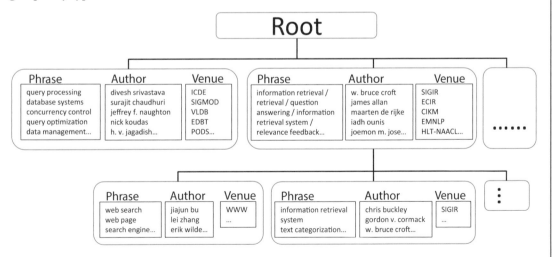

Figure 2.5: Sample output from CATHYHIN—topical hierarchy of multi-typed entities. Each node has a ranked list of phrases and two ranked entity lists.

2.2.1   THE BASIC MODEL

We first introduce the basic generative model, which considers all link types to be equally important. For a given topic t, we assume $C_t = k$. The value of k can be either specified by users or chosen using a model selection criterion. We discuss the choice of k in Section 2.2.3.

We assume every link has a direction. For undirected networks, we convert them to directed networks by duplicating each link between v_i^x and v_j^y in both directions $v_i^x \rightarrow v_j^y$ and $v_j^y \rightarrow v_i^x$. So this model can be applied to both undirected and directed networks.

To derive the model, we first assume the links between any two nodes can be decomposed into one or multiple unit-weight links (e.g., a link with weight 2 can be seen as a summation of two unit-weight links). Later we will discuss the case where the link weight is not an integer. Each unit-weight link has a topic label, which is either a subtopic $t \odot z, z \in [k]$, or a dummy label $t \odot 0$, implying the link is generated by a background topic and should not be attributed to any subtopic of t.

Table 2.1: Notations used in CATHYHIN

Symbol	Description
G^t	the edge-weighted network associated with topic t
$\mathcal{V}_x^t$	the set of nodes of type x in topic t
$\mathcal{E}_{x,y}^t$	the set of non-zero link weights of type (x, y) in topic t
π_t	the parent topic of topic t
C_t	the number of child topics of topic t
o	the root topic
m	the number of node types
l	the number of link types
$n_{x,y}$	the total number of type-x and type-y node pairs that have links
v_i^x	the i-th node of type x
$e_{i,j}^{x,y,t}$	the link weight between v_i^x and v_j^y in topic t
M_t	the sum of link weight in topic t: $\sum_{i,j,x,y} e_{i,j}^{x,y,t}$
$M_t^{x,y}$	the sum of type-(x, y) link weight in topic t: $\sum_{i,j} e_{i,j}^{x,y,t}$
ϕ_t^x	the distribution over type-x nodes in topic t
$\phi_{t/0}^x$	the background distribution over type-x nodes in topic t
ρ	the distribution over subtopics
θ	the distribution over link types
$\alpha_{x,y}$	the importance of link type (x, y)

The generative process for a link with unit weight is as follows.

1. Generate the topic label z according to a multinomial distribution ρ_t.

2. Generate the link type (x, y) according to a multinomial distribution θ.

3. If $z \in [k]$:

    (a) Generate the first end node u_1 from the type-x ranking distribution $\phi_{t \odot z}^x$.

(b) Generate the second end node u_2 from the type-y ranking distribution $\phi_{t \odot z}^y$.

Otherwise, ($z = 0$):

(a) Generate the first end node u_1 from the type-x ranking distribution $\phi_{t \odot 0}^x$.

(b) Generate the second end node u_2 from the type-y ranking distribution ϕ_t^y.

Example 2.5 A link between two terms *query* and *processing* in a topic $t \odot z = t \odot 1$ (*Database*) may be generated in the following order: (i) generate the topic label $z = 1$ with probability $\rho_{t \odot z} = 0.2$; (ii) generate the link type $(x, y) = (term, term)$ according to $\theta_{x,y} = 0.15$; (iii) generate the first end node $u_1 = query$ from the term distribution $\phi_{t \odot z, u_1}^x = 0.004$; and (iv) generate the second end node $u_2 = processing$ from the term distribution $\phi_{t \odot z, u_2}^y = 0.001$.

This process is repeated for M^t times, to generate all the unit-weight links. Note that when generating a background topic link, the two nodes i and j are not symmetric. The first end node is a background node, and can have a background topic link with any other nodes based simply on their importance in the parent topic, irrespective of any subtopic. Highly ranked nodes in the background topic tend to have a link distribution over all nodes that is similar to their distribution in the parent topic. This part can be altered if the network follows a different assumption about the background nodes. See Figure 2.6 for a graphical representation of the model.

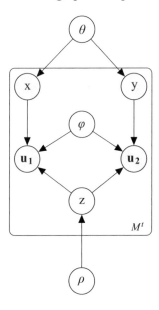

Figure 2.6: The generative process of the "unit-weight" links.

With these generative assumptions for each unit-weight link, we can derive the distribution of link weight for any two nodes (v_i^x, v_j^y). First, we notice that the total number of topic-$t \odot z$

unit-weight links is expected to be $M^t \rho_{t \odot z}$. It implies that we are expected to repeat the generation of topic-$t \odot z$ unit-weight links for $M^t \rho_{t \odot z}$ times. Second, we investigate how many topic-$t \odot z$ unit-weight links will be generated between two certain nodes v_i^x and v_j^y, among all the topic-$t \odot z$ links. Consider the event, the two end nodes are v_i^x and v_j^y, for each generated topic-$t \odot z$ link. It is a Bernoulli trial with success probability $\theta_{x,y} \phi_{t \odot z,i}^x \phi_{t \odot z,j}^y$ for $z \in [k]$. When $M^t \rho_{t \odot z}$ is large, the total number of successes $e_{i,j}^{x,y,t \odot z}$ asymptotically follows a Poisson distribution $Pois\left(M^t \rho_{t \odot z} \theta_{x,y} \phi_{t \odot z,i}^x \phi_{t \odot z,j}^y\right)$. Similarly, the total number of background topic links $e_{i,j}^{x,y,t \odot 0}$ asymptotically follows a Poisson distribution $Pois\left(M^t \rho_{t \odot 0} \theta_{x,y} \phi_{t \odot 0,i}^x \phi_{t,j}^y\right)$.

One important implication due to the *additive* property of Poisson distribution is:

$$e_{i,j}^{x,y,t} = \sum_{z=0}^{k} e_{i,j}^{x,y,t \odot z} \sim Poisson\left(M^t \theta_{x,y} s_{i,j}^{x,y,t}\right), \tag{2.8}$$

where $s_{i,j}^{x,y,t} = \sum_{z=1}^{k} \rho_{t \odot z} \phi_{t \odot z,i}^x \phi_{t \odot z,j}^y + \rho_{t \odot 0} \phi_{t \odot 0,i}^x \phi_{t,j}^y$.

This leads to a "collapsed" model as depicted in Figure 2.7. Though we have so far assumed the link weight to be an integer, this collapsed model remains valid with non-integer link weights (due to Lemma 2.7, discussed in Section 2.2.2).

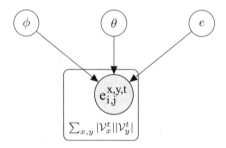

Figure 2.7: The "collapsed" generative process of the link weights.

Given the model parameters, the probabilities of all observed links are:

$$p\left(\{e_{i,j}^{x,y,t}\}|\theta,\rho,\phi\right) = \prod_{v_i^x, v_j^y} \frac{\left(M^t \theta_{x,y} s_{i,j}^{x,y,t}\right)^{e_{i,j}^{x,y,t}} \exp\left(-M^t \theta_{x,y} s_{i,j}^{x,y,t}\right)}{e_{i,j}^{x,y,t}!}. \tag{2.9}$$

We learn the parameters by the *Maximum Likelihood* (ML) principle: Find the parameter values that maximize the likelihood in Eq. (2.9). First, we take the logarithm and remove the part independent of the parameters:

$$L(\theta, \rho, \phi) = \sum_{v_i^x, v_j^y} \left[e_{i,j}^{x,y,t} \log\left(\theta_{x,y} s_{i,j}^{x,y,t}\right) - M^t \theta_{x,y} s_{i,j}^{x,y,t} \right]$$

$$= \sum_{v_i^x, v_j^y} e_{i,j}^{x,y,t} \left[\log \theta_{x,y} + \log\left(\sum_{z=1}^{k} \rho_{t\odot z} \phi_{t\odot z,i}^x \phi_{t\odot z,j}^y + \rho_{t\odot 0} \phi_{t\odot 0,i}^x \phi_{t,j}^y \right) \right]$$

$$- M^t \sum_{v_i^x, v_j^y} \theta_{x,y} \left(\sum_{z=1}^{k} \rho_{t\odot z} \phi_{t\odot z,i}^x \phi_{t\odot z,j}^y + \rho_{t\odot 0} \phi_{t\odot 0,i}^x \phi_{t,j}^y \right).$$

The gradient method is hard to apply due to the logarithm of summation. We introduce an auxiliary distribution $q_{i,j}^{x,y}$ over subtopics for every link (v_i^x, v_j^y). They satisfy

$$\sum_{z=0}^{k} q_{i,j}^{x,y,z} = 1, \forall v_i^x, v_j^y. \tag{2.10}$$

Due to Jensen's inequality, we have:

$$\log\left(\sum_{z=1}^{k} \rho_{t\odot z} \phi_{t\odot z,i}^x \phi_{t\odot z,j}^y + \rho_{t\odot 0} \phi_{t\odot 0,i}^x \phi_{t,j}^y \right)$$

$$= \log\left(\sum_{z=1}^{k} q_{i,j}^{x,y,z} \frac{\rho_{t\odot z} \phi_{t\odot z,i}^x \phi_{t\odot z,j}^y}{q_{i,j}^{x,y,z}} + q_{i,j}^{x,y,0} \frac{\rho_{t\odot 0} \phi_{t\odot 0,i}^x \phi_{t,j}^y}{q_{i,j}^{x,y,0}} \right)$$

$$\geq \sum_{z=1}^{k} q_{i,j}^{x,y,z} \log \frac{\rho_{t\odot z} \phi_{t\odot z,i}^x \phi_{t\odot z,j}^y}{q_{i,j}^{x,y,z}} + q_{i,j}^{x,y,0} \log \frac{\rho_{t\odot 0} \phi_{t\odot 0,i}^x \phi_{t,j}^y}{q_{i,j}^{x,y,0}}$$

where the equality holds if and only if:

$$\frac{\rho_{t\odot z} \phi_{t\odot z,i}^x \phi_{t\odot z,j}^y}{q_{i,j}^{x,y,z}} = \frac{\rho_{t\odot 0} \phi_{t\odot 0,i}^x \phi_{t\odot j}^y}{q_{i,j}^{x,y,0}}, \forall z = 1, \ldots, k. \tag{2.11}$$

We define an auxiliary function $F(q, \theta, \rho, \phi)$:

$$F(q, \theta, \rho, \phi) =$$

$$\sum_{v_i^x, v_j^y} e_{i,j}^{x,y,t} \left(\log \theta_{x,y} + \sum_{z=1}^{k} q_{i,j}^{x,y,z} \log \frac{\rho_{t\odot z} \phi_{t\odot z,i}^x \phi_{t\odot z,j}^y}{q_{i,j}^{x,y,z}} + q_{i,j}^{x,y,0} \log \frac{\rho_{t\odot 0} \phi_{t\odot 0,i}^x \phi_{t,j}^y}{q_{i,j}^{x,y,0}} \right)$$

$$- M^t \sum_{v_i^x, v_j^y} \theta_{x,y} \left(\sum_{z=1}^{k} \rho_{t\odot z} \phi_{t\odot z,i}^x \phi_{t\odot z,j}^y + \rho_{t\odot 0} \phi_{t\odot 0,i}^x \phi_{t,j}^y \right).$$

F can be maximized by iteratively applying two alternating steps: (i) Fix θ, ρ, ϕ, choose q to optimize F; and (ii) fix q, choose θ, ρ, ϕ to optimize F. For (i), Eq. (2.11) and Eq. (2.10) together imply:

$$q_{i,j}^{x,y,z}(\theta,\rho,\phi) = \frac{\rho_{t\odot z}\phi_{t\odot z,i}^x\phi_{t\odot z,j}^y}{\sum_{c=1}^k \rho_{t\odot c}\phi_{t\odot c,i}^x\phi_{t\odot c,j}^y + \rho_{t\odot 0}\phi_{t\odot 0,i}^x\phi_{t,j}^y}, z = 1,\ldots,k \qquad (2.12)$$

$$q_{i,j}^{x,y,0}(\theta,\rho,\phi) = \frac{\rho_{t\odot 0}\phi_{t\odot 0,i}^x\phi_{t,j}^y}{\sum_{c=1}^k \rho_{t\odot c}\phi_{t\odot c,i}^x\phi_{t\odot c,j}^y + \rho_{t\odot 0}\phi_{t\odot 0,i}^x\phi_{t,j}^y}. \qquad (2.13)$$

For (ii), we use the Lagrange multiplier method to incorporate the probability constraints for $\theta, \rho,$ and ϕ. The gradient method yields a closed-form solution:

$$\theta_{x,y}(q) = \frac{\sum_{v_i^x,v_j^y} e_{i,j}^{x,y,t}}{M^t} \qquad (2.14)$$

$$\rho_{t,z}(q) = \frac{\sum_{v_i^x,v_j^y} e_{i,j}^{x,y,t} q_{i,j}^{x,y,z}}{M^t} \qquad (2.15)$$

$$\phi_{t/z,i}^x(q) = \frac{\sum_{v_j^y}(e_{i,j}^{x,y,t} q_{i,j}^{x,y,z} + e_{j,i}^{y,x,t} q_{j,i}^{y,x,z})}{\sum_{v_u^x,v_j^y}(e_{u,j}^{x,y,t} q_{u,j}^{x,y,z} + e_{j,u}^{y,x,t} q_{j,u}^{y,x,z})}, z = 1,\ldots,k \qquad (2.16)$$

$$\phi_{t/0,i}^x(q) = \frac{\sum_{v_j^y} e_{i,j}^{x,y,t} q_{i,j}^{x,y,0}}{\sum_{v_u^x,v_j^y} e_{u,j}^{x,y,t} q_{u,j}^{x,y,0}}. \qquad (2.17)$$

During the iterations, the value of function F keeps non-decreasing. Let $q^{(a)}, \theta^{(a)}, \rho^{(a)}, \phi^{(a)}$ denote the parameter values after the a-th iteration. We then have: $L(\theta^{(a)}, \rho^{(a)}, \phi^{(a)}) = F(q^{(a+1)}, \theta^{(a)}, \rho^{(a)}, \phi^{(a)})$. Therefore, the value of $L(\theta^{(a)}, \rho^{(a)}, \phi^{(a)})$ also keeps non-decreasing during the iterations. Since the function L is upper bounded, $L(\theta^{(a)}, \rho^{(a)}, \phi^{(a)})$ eventually converges to a local maximum.

This solution is similar to an Expectation-Maximization (EM) algorithm that is used for ML inference for many statistical models. In fact, we have the following theorem.

Theorem 2.6   *The solution Eqs. (2.12)–(2.17) derived from the collapsed model (Figure 2.6) is equivalent to an EM solution derived from the unrolled model (Figure 2.7).*

Proof. In the unrolled model, the likelihood of a unit-weight link can be written as:

$$p(x, y, u_1, u_2 | \theta, \rho, \phi, t) = p(x, y | \theta) \sum_z p(z | \rho_t) p(u_1 | x, t \odot z, \phi) p(u_2 | y, t \odot z, \phi)$$

in which the topic z is a latent variable. The EM algorithm iteratively applies the following two steps.

Expectation step (E-step). Calculate the expected value of the log likelihood function with respect to the conditional distribution of latent variables given observed variables under the current estimate of the parameters $\theta^{(a)}, \rho^{(a)}, \phi^{(a)}$:

$$Q(\theta, \rho, \phi | \theta^{(a)}, \rho^{(a)}, \phi^{(a)})$$
$$= \sum_{x,y,u_1,u_2} \sum_{z} p(z|x, y, t, u_1, u_2, \theta^{(a)}, \rho^{(a)}, \phi^{(a)}) \log p(x, y, u_1, u_2 | t \odot z, \theta, \rho, \phi)$$
$$= \sum_{x,y,u_1,u_2} \sum_{z} p(z|x, y, t, u_1, u_2, \theta^{(a)}, \rho^{(a)}, \phi^{(a)}) \log p(x, y|\theta) p(u_1 | t \odot z, \phi) p(u_2 | t \odot z, \phi).$$

Applying Bayes' theorem, we have:

$$p(z|x, y, u_1, u_2, \theta^{(a)}, \rho^{(a)}, \phi^{(a)}, t) = \frac{p(z, x, y, u_1, u_2 | \theta^{(a)}, \rho^{(a)}, \phi^{(a)}, t)}{p(x, y, u_1, u_2 | \theta^{(a)}, \rho^{(a)}, \phi^{(a)}, t)}$$
$$= \frac{p(x, y|\theta^{(a)}) p(z|\rho_t^{(a)}) p(u_1 | t \odot z, \phi^{(a)}) p(u_2 | t \odot z, \phi^{(a)})}{\sum_c p(x, y|\theta^{(a)}) p(c|\rho_t^{(a)}) p(u_1 | t \odot c, \phi^{(a)}) p(u_2 | t \odot c, \phi^{(a)})}$$
$$= \frac{p(z|\rho_t^{(a)}) p(u_1 | t \odot z, \phi^{(a)}) p(u_2 | t \odot z, \phi^{(a)})}{\sum_c p(c|\rho_t^{(a)}) p(u_1 | t \odot c, \phi^{(a)}) p(u_2 | t \odot c, \phi^{(a)})}$$
$$= \begin{cases} \dfrac{\rho_{t \odot z} \phi^x_{t \odot z, u_1} \phi^y_{t \odot z, u_2}}{\sum_{c=1}^k \rho_{t \odot c} \phi^x_{t \odot c, u_1} \phi^y_{t \odot c, u_2} + \rho_{t \odot 0} \phi^x_{t \odot 0, u_1} \phi^y_{t, u_2}} & z \in [k] \\[2ex] \dfrac{\rho_{t \odot 0} \phi^x_{t \odot 0, u_1} \phi^y_{t, u_2}}{\sum_{c=1}^k \rho_{t \odot c} \phi^x_{t \odot c, u_1} \phi^y_{t \odot c, u_2} + \rho_{t \odot 0} \phi^x_{t \odot 0, u_1} \phi^y_{t \odot u_2}} & z = 0. \end{cases} \tag{2.18}$$

We omit the superscript (a) in Eq. (2.18). Comparing Eq. (2.18) with Eqs. (2.12) and (2.13), we find that the auxiliary distribution q we introduced above is actually equal to the posterior distribution over the topics on each link. Now we can write Q as:

$$Q(\theta, \rho, \phi | \theta^{(a)}, \rho^{(a)}, \phi^{(a)})$$
$$= \sum_{x,y,u_1,u_2} \sum_{z} p(z|x, y, u_1, u_2, \theta^{(a)}, \rho^{(a)}, \phi^{(a)}, t) \log p(x, y|\theta) p(u_1 | t \odot z, \phi) p(u_2 | t \odot z, \phi)$$
$$= \sum_{x,y,e^{x,y,t}_{i,j} \in \mathcal{E}^t_{x,y}} e^{x,y,t}_{i,j} \left(\sum_{z=1}^{k} q^{x,y,z(a)}_{i,j} \log \phi^x_{t \odot z, i} \phi^y_{t \odot z, j} + q^{x,y,0(a)}_{i,j} \log \phi^x_{t \odot 0, i} \phi^y_{t, j} + \log \theta_{x,y} \right).$$

Maximization step (M-step). Find the parameters that maximize Q:

$$\theta^{(a+1)}, \rho^{(a+1)}, \phi^{(a+1)} = \arg \max_{\theta, \rho, \phi} Q(\theta, \rho, \phi | \theta^{(a)}, \rho^{(a)}, \phi^{(a)}).$$

Using the Lagrange multiplier method, we can obtain the solution:

$$\theta_{x,y}^{(a+1)} = \frac{\sum_{e_{i,j}^{x,y,t} \in \mathcal{E}_{x,y}^t} e_{i,j}^{x,y,t}}{M^t} \tag{2.19}$$

$$\rho_{t\odot z}^{(a+1)} = \frac{\sum_{e_{i,j}^{x,y,t} \in \mathcal{E}_{x,y}^t} e_{i,j}^{x,y,t} q_{i,j}^{x,y,z\,(a)}}{M^t} \tag{2.20}$$

$$\phi_{t\odot z,i}^{x\ (a+1)} = \frac{\sum_{e_{i,j}^{x,y,t} \in \mathcal{E}_{x,y}^t} e_{i,j}^{x,y,t} q_{i,j}^{x,y,z\,(a)} + \sum_{e_{j,i}^{y,x,t} \in \mathcal{E}_{y,x}^t} e_{j,i}^{y,x,t} q_{j,i}^{y,x,z\,(a)}}{\sum_{e_{u,j}^{x,y,t} \in \mathcal{E}_{x,y}^t} e_{u,j}^{x,y,t} q_{u,j}^{x,y,z\,(a)} + \sum_{e_{j,u}^{y,x,t} \in \mathcal{E}_{y,x}^t} e_{j,u}^{y,x,t} q_{j,u}^{y,x,z\,(a)}} \tag{2.21}$$

$$\phi_{t\odot 0,i}^{x\ (a+1)} = \frac{\sum_{e_{i,j}^{x,y,t} \in \mathcal{E}_{x,y}^t} e_{i,j}^{x,y,t} q_{i,j}^{x,y,0\,(a)}}{\sum_{e_{u,j}^{x,y,t} \in \mathcal{E}_{x,y}^t} e_{u,j}^{x,y,t} q_{u,j}^{x,y,0\,(a)}}. \tag{2.22}$$

It is easy to verify the equivalence of Eqs. (2.19)–(2.22) and Eqs. (2.14)–(2.17). ■

This theorem shows that we can derive the same solution from both the unrolled generative model and the collapsed model. The unrolled model is natural and intuitive, but the collapsed model is easier for extension, as we will see in the next subsection.

The theorem also incarnates q as a posterior distribution over the topics on each link. Based on this, we can calculate the expected number of topic-$t \odot z$ links between every two nodes:

$$\hat{e}_{i,j}^{x,y,t\odot z} = e_{i,j}^{x,y,t} q_{i,j}^{x,y,z}. \tag{2.23}$$

Then we have the update rules based on $\hat{e}_{i,j}^{x,y,z}$.

• E-step:

$$\hat{e}_{i,j}^{x,y,t\odot z} = \frac{e_{i,j}^{x,y,t} \rho_{t\odot z} \phi_{t\odot z,i}^x \phi_{t\odot z,j}^y}{\sum_{c=1}^k \rho_{t,c} \phi_{t\odot c,i}^x \phi_{t\odot c,j}^y + \rho_{t\odot 0} \phi_{t\odot 0,i}^x \phi_{t,j}^y} \tag{2.24}$$

$$\hat{e}_{i,j}^{x,y,t\odot 0} = \frac{e_{i,j}^{x,y,t} \rho_{t\odot 0} \phi_{t\odot 0,i}^x \phi_{t,j}^y}{\sum_{c=1}^k \rho_{t\odot c} \phi_{t\odot c,i}^x \phi_{t\odot c,j}^y + \rho_{t\odot 0} \phi_{t\odot 0,i}^x \phi_{t,j}^y}. \tag{2.25}$$

- M-step:

$$\theta_{x,y} = \frac{\sum_{v_i^x, v_j^y} e_{i,j}^{x,y,t}}{M^t} \tag{2.26}$$

$$\rho_{t\odot z} = \sum_{v_i^x, v_j^y} \frac{\hat{e}_{i,j}^{x,y,t\odot z}}{M^t} \tag{2.27}$$

$$\phi_{t\odot z,i}^x = \frac{\sum_{v_j^y} (\hat{e}_{i,j\cdot}^{x,y,t\odot z} + \hat{e}_{j,i}^{y,x,t\odot z})}{\sum_{v_u^x, v_j^y} (\hat{e}_{u,j}^{x,y,t\odot z} + \hat{e}_{j,u}^{y,x,t\odot z})} \tag{2.28}$$

$$\phi_{t\odot 0,i}^x = \frac{\sum_{v_j^y} \hat{e}_{i,j}^{x,y,t\odot 0}}{\sum_{v_u^x, v_j^y} \hat{e}_{u,j}^{x,y,t\odot 0}}. \tag{2.29}$$

These equations are intuitive. In the E-step, the expected link weight of each subtopic $\hat{e}$ is calculated from the posterior distribution q given the current parameter estimates. This can be viewed as soft clustering of links. In the M-step, the parameters are re-estimated based on the link clustering: the link type weight $\theta_{x,y}$ is calculated as dividing the total link weight of type (x, y) by the total link weight; the topic distribution ρ_t is estimated by the expected number of links in each subtopic; and the ranking distribution over nodes in each topic $t \odot z$ is estimated by the total number of topic-$t \odot z$ links associated these nodes.

We update $\hat{e}, \phi, \rho$ at each iteration because $\theta_{x,y}$ is a constant. The EM algorithm can be executed multiple times with random initializations, and the solution with the best likelihood will be chosen.

The subnetwork for topic $t \odot z$ is naturally extracted from the estimated $\hat{e}$ (expected link weight attributed to each topic). For efficiency purposes, we remove links whose weight is less than 1, and then filter out all resulting isolated nodes. We can then recursively apply the same generative model to the constructed subnetworks until the desired hierarchy is constructed.

2.2.2 LEARNING LINK-TYPE WEIGHTS

The generative model described above does not differentiate between the importance of different link types. However, we may wish to discover topics that are biased toward certain types of links, and the bias may vary at different levels of the hierarchy. For example, in the computer science domain, the links between venues and other entities may be more important indicators than other link types in the top level of the hierarchy; however, these same links may be less useful for discovering sub-areas in the lower levels (e.g., authors working in different sub-areas may publish in the same venue).

We therefore extend the basic model to capture the importance of different link types. We introduce a *link-type weight* $\alpha_{x,y} > 0$ for each link type (x, y). We use these weights to scale a link's observed weight up or down, so that a unit-weight link of type (x, y) in the original network will have a *scaled* weight $\alpha_{x,y}$. Thus, a link of type (x, y) is valued more when $\alpha_{x,y} > 1$, less when $0 < \alpha_{x,y} < 1$, and becomes negligible as $\alpha_{x,y}$ approaches 0.

When the link type weights $\alpha_{x,y}$ are specified for the model, the EM inference algorithm is unchanged, with the exception that all the $e_{i,j}^{x,y,t}$ in the update equations should be replaced by $\alpha_{x,y} e_{i,j}^{x,y,t}$. When all $\alpha_{x,y}$'s are equal, the weight-learning model reduces to the basic model. Most of the time, the weights of the link types will not be specified explicitly by users, and must therefore be learned from the data.

We first note an important property of the model, justifying our previous claim that link weights need not be integers.

Lemma 2.7 Scale-invariant *The EM solution is invariant to a constant scale-up of all the link weights. That is, if we replace all the $e_{i,j}^{x,y,t}$ with $c e_{i,j}^{x,y,t}$, all the resulting $q_{i,j}^{x,y,z}$, $\rho_{t \odot z}$, $\theta_{x,y}$ and $\phi_{t \odot z,i}^{x}$ remain unchanged for topic t as well as for all the descendant topics of t.*

The proof is straightforward by induction.

With the scale-invariant property on the link weights, we can prove the following theorem.

Theorem 2.8 *For a set of l positive numbers $\alpha_{x,y} > 0$, there exist another set of l positive numbers $\beta_{x,y} > 0$, such that the EM solution based on link weights $\alpha_{x,y}$ and $\beta_{x,y}$ are identical, and $\prod_{e_{i,j}^{x,y,t}>0} e_{i,j}^{x,y,t} = \prod_{e_{i,j}^{x,y,t}>0} (\beta_{x,y} e_{i,j}^{x,y,t})$.*

Proof: Let $\chi = \dfrac{\prod_{e_{i,j}^{x,y,t}>0} e_{i,j}^{x,y,t}}{\prod_{e_{i,j}^{x,y,t}>0} (\alpha_{x,y} e_{i,j}^{x,y,t})}$, $N = \sum_{x,y} n_{x,y}$. We define:

$$\beta_{x,y} \equiv \chi^{\frac{1}{N}} \alpha_{x,y}. \tag{2.30}$$

The scale-invariant property implies that the EM solution based on link weights $\alpha_{x,y}$ and $\beta_{x,y}$ are identical. So we have:

$$\prod_{e_{i,j}^{x,y,t}>0} (\beta_{x,y} e_{i,j}^{x,y,t}) = \prod_{e_{i,j}^{x,y,t}>0} (\chi^{\frac{1}{N}} \alpha_{x,y} e_{i,j}^{x,y,t})$$
$$= \chi \prod_{e_{i,j}^{x,y,t}>0} (\alpha_{x,y} e_{i,j}^{x,y,t}) = \prod_{e_{i,j}^{x,y,t}>0} e_{i,j}^{x,y,t}. \tag{2.31}$$

∎

With this theorem, we can assume that *w.l.o.g.*, the product of all the non-zero link weights remains invariant before and after scaling:

$$\prod_{e_{i,j}^{x,y,t}>0} e_{i,j}^{x,y,t} = \prod_{e_{i,j}^{x,y,t}>0} (\alpha_{x,y} e_{i,j}^{x,y,t}) \tag{2.32}$$

which reduces to $\prod_{x,y} \alpha_{x,y}^{n_{x,y}} = 1$, where $n_{x,y} = |\mathcal{E}_{x,y}^t|$ is the number of non-zero links with type (x, y). With this constraint, we maximize the likelihood $p(\{e_{i,j}^{x,y,t}\}|\theta, \rho, \phi, \alpha)$:

$$\max_{v_i^x, v_j^y} \prod \frac{(\alpha_{x,y} M_{x,y}^t s_{i,j}^{x,y,t})^{\alpha_{x,y} e_{i,j}^{x,y,t}} \exp(-\alpha_{x,y} M_{x,y}^t s_{i,j}^{x,y,t})}{(\alpha_{x,y} e_{i,j}^{x,y,t})!} \tag{2.33}$$

$$s.t. \prod_{x,y} \alpha_{x,y}^{n_{x,y}} = 1, \alpha_{x,y} > 0, \tag{2.34}$$

where $M_{x,y}^t = \sum_{v_i^x, v_j^y} e_{i,j}^{x,y,t}$ is the total weight for type (x, y) links. With Sterling's approximation $n! \sim (\frac{n}{e})^n \sqrt{2\pi n}$, we rewrite the log likelihood:

$$\max_{v_i^x, v_j^y} \sum \left(\alpha_{x,y} e_{i,j}^{x,y,t} \log(\alpha_{x,y} M_{x,y}^t s_{i,j}^{x,y,t}) - \alpha_{x,y} M_{x,y}^t s_{i,j}^{x,y,t} \right. \tag{2.35}$$

$$\left. -\alpha_{x,y} e_{i,j}^{x,y,t} [\log(\alpha_{x,y} e_{i,j}^{x,y,t}) - 1] - \frac{1}{2} \log(\alpha_{x,y} e_{i,j}^{x,y,t}) \right)$$

$$s.t. \sum_{x,y} n_{x,y} \log \alpha_{x,y} = 0. \tag{2.36}$$

Using the Lagrange multiplier method, we can find the optimal value for α when the other parameters are fixed:

$$\alpha_{x,y} = \frac{\left[\prod_{x,y} \left(\frac{1}{n_{x,y}} \sum_{i,j} e_{i,j}^{x,y,t} \log \frac{e_{i,j}^{x,y,t}}{M_{x,y}^t s_{i,j}^{x,y,t}} \right)^{n_{x,y}} \right]^{\frac{1}{\sum_{x,y} n_{x,y}}}}{\frac{1}{n_{x,y}} \sum_{i,j} e_{i,j}^{x,y,t} \log \frac{e_{i,j}^{x,y,t}}{M_{x,y}^t s_{i,j}^{x,y,t}}}. \tag{2.37}$$

With some transformation of the denominator:

$$\sigma_{x,y} = \frac{1}{n_{x,y}} \sum_{i,j} e_{i,j}^{x,y,t} \log \frac{e_{i,j}^{x,y,t}}{M_{x,y}^t s_{i,j}^{x,y,t}} \tag{2.38}$$

$$= \frac{M_{x,y}^t}{n_{x,y}} \sum_{v_i^x, v_j^y} \frac{e_{i,j}^{x,y,t}}{M_{x,y}^t} \log \frac{e_{i,j}^{x,y,t} / M_{x,y}^t}{s_{i,j}^{x,y,t}}$$

we can see more clearly that the link type weight is negatively correlated with two factors: the average link weight $\frac{M_{x,y}^t}{n_{x,y}}$ and the KL-divergence of the expected link weight distribution to the observed link weight distribution $\sum_{v_i^x, v_j^y} \frac{e_{i,j}^{x,y,t}}{M_{x,y}^t} \log \frac{e_{i,j}^{x,y,t} / M_{x,y}^t}{s_{i,j}^{x,y,t}}$. The first factor is used to balance the scale of link weights of different types (e.g., a type-1 link always has X times greater weight than a type-2 link). The second factor measures the importance of a link type in the model. The more the prediction diverges from the observation, the worse the quality of a link type.

Therefore, we have the following iterative algorithm for optimizing the joint likelihood.

1. Initialize all the parameters.

2. Fixing α, update ρ, ϕ using EM equations:

$$\hat{e}_{i,j}^{x,y,z} = \frac{\alpha_{x,y} e_{i,j}^{x,y,t} \rho_{t \odot z} \phi_{t \odot z,i}^{x} \phi_{t \odot z,j}^{y}}{\sum_{c=1}^{k} \rho_{t \odot c} \phi_{t \odot c,i}^{x} \phi_{t \odot c,j}^{y} + \rho_{t \odot 0} \phi_{t \odot 0,i}^{x} \phi_{t,j}^{y}} \tag{2.39}$$

$$\hat{e}_{i,j}^{x,y,t \odot 0} = \frac{\alpha_{x,y} e_{i,j}^{x,y,t} \rho_{t \odot 0} \phi_{t \odot 0,i}^{x} \phi_{t,j}^{y}}{\sum_{c=1}^{k} \rho_{t \odot c} \phi_{t \odot c,i}^{x} \phi_{t \odot c,j}^{y} + \rho_{t \odot 0} \phi_{t \odot 0,i}^{x} \phi_{t,j}^{y}} \tag{2.40}$$

$$\rho_{t \odot z} = \frac{\sum_{v_i^x, v_j^y} \hat{e}_{i,j}^{x,y,t \odot z}}{\sum_{x,y} \alpha_{x,y} M_{x,y}^{t}} \tag{2.41}$$

$$\phi_{t \odot z,i}^{x} = \frac{\sum_{v_j^y} (\hat{e}_{i,j}^{x,y,t \odot z} + \hat{e}_{j,i}^{y,x,t \odot z})}{\sum_{v_u^x, v_j^y} (\hat{e}_{u,j}^{x,y,t \odot z} + \hat{e}_{j,u}^{y,x,t \odot z})} \tag{2.42}$$

$$\phi_{t \odot 0,i}^{x} = \frac{\sum_{v_j^y} \hat{e}_{i,j}^{x,y,t \odot 0}}{\sum_{v_u^x, v_j^y} \hat{e}_{u,j}^{x,y,t \odot 0}}. \tag{2.43}$$

3. Fixing ρ, ϕ, update α using Eq. (2.37).

4. Repeat Steps 2 and 3 until the likelihood converges.

For each iteration, the time complexity is $\mathcal{O}(\sum_{x,y} n_{x,y})$, that is, linear to the total number of non-zero links. The likelihood is guaranteed to converge to a local optimum. Once again, a random initialization strategy can be employed to choose a solution with the best local optimum.

It is important to note that the learned link weights indicate the overall importance of a type in the following sense: how much the original link weight of each type should be rescaled. A larger link weight corresponds to a link type that should be counted more when we fit the likelihood of the observation. If we want to see the subtle difference of the importance for each individual link, a possible modeling strategy is to parameterize the link weights $\alpha_{x,y}$ (e.g., according to the attributes or topological features of nodes such as their degree or weighted degree).

2.2.3 SHAPE OF HIERARCHY

In this section, we discuss the following issues that affect the shape of the constructed hierarchy.

- **Number of children for each topic.** Since our framework is recursive, the shape of the tree is essentially determined by how many children each node has, that is, how many subtopics each topic has. For every topic, our model can work with an arbitrary number of subtopics that is larger than 1. However, it may be more reasonable to have a certain number of subtopics than others. In general, we prefer each topic to have a small number of subtopics (e.g., between 2 and 10) to make it easy for browsing. For example, if the root has 5 subtopics and each of them has 4 subtopics, the 3 level hierarchy is in general easier to browse than directly showing all 20 topics under the root.

Given a range of the numbers of subtopics, such as [2, 10], we would like to choose a reasonable number of children k for each topic. It is a model selection problem. Among various model selection strategies in the literature, we select two of them and introduce how they can be adapted for our model.

The first strategy was proposed by Smyth [2000] to adopt cross-validation to choose the parameter K. In our setting, we can first fit the generative model to a sampled subnetwork H^t of the given network G^t. Then we evaluate the likelihood of the model on the rest part of the network $G^t - H^t$, which is called the held-out network. By checking the averaged held-out likelihood with varying number of sub-clusters, the parameter with the maximum value will be chosen as the best candidate.

The second strategy is based on the Bayesian information criterion (BIC). A similar criterion is Akaike information criterion (AIC). Both BIC and AIC resolve the overfitting problem. When we increase the number of topics k, it is possible to increase the likelihood, but may result in overfitting because the model will have a larger number of parameters. BIC and AIC introduce a penalty term for the number of parameters in the model, and the penalty term is larger in BIC than in AIC. Using BIC, the measure for our model is defined as:

$$BIC = -2 \log p \left(\{e_{i,j}^{x,y,t}\} | \theta, \rho, \phi \right) + |\theta, \rho, \phi| \cdot \log |\mathcal{E}^t|,$$

where $|\theta, \rho, \phi|$ is the number of free parameters in the model and $|\mathcal{E}^t|$ refers to the size of observed links. As we only care about the number of topics k, $|\theta, \rho, \phi|$ can be reduced to $|\mathcal{V}^t| k$ plus a constant independent of k, where $|\mathcal{V}^t|$ is the number of nodes. We can then select k with the largest BIC score.

BIC is derived under the assumption that the data distribution is in the exponential family. Cross validation only assumes that the sampled network and the held-out network are generated from the same model. Comparing these two criteria, we generally recommend cross validation over BIC when there are sufficient data. However, when the network is small, cross validation is prone to high variation and BIC can be used as an alternative.

- **Depth of the hierarchy.** A simple and intuitive strategy to decide the depth of the hierarchy is to rely on the selected number of children mentioned above. For example, if the best number of topics is $k = 1$, it implies we should stop expanding the current topic node. In practice, we can set a threshold on the largest depth of the tree, as well as the size of the network. Once the tree has reached the maximal depth, or the size of the network in current topic is too small, we can terminate the recursion. A general implication is that the more children each node has, the less deep the final hierarchy will be.

- **Balance of subtree size.** The distribution of ρ_t's determines the size of subtrees. The more evenly distributed ρ_t's are, the more balanced the subtrees. Generally, we would like to generate a balanced tree because it is efficient for browsing. If this is the case, we should randomly initialize the topic of each link from a uniform distribution. In case one would like to generate a skewed hierarchy, the random initialization of each link's topic distribution should follow a non-uniform multinomial, which can be generated from a Dirichlet prior. Our model can be extended into a Bayesian framework, which can incorporate conjugate prior for all the parameters. The shape of the hierarchy can then be controled by the hyperparameters of the prior.

2.3   EMPIRICAL ANALYSIS

Table 2.2: # Links in CATHYHIN datasets

DBLP (# Nodes)	**Term (6,998)**	**Author (12,886)**	**Venue (20)**
Term	693,132	900,201	104,577
Author	–	156,255	99,249
NEWS (# Nodes)	**Term (13,129)**	**Person (4,555)**	**Location (3,845)**
Term	686,007	386,565	506,526
Person	–	53,094	129,945
Location	–	–	85,047

The lack of gold standard is a known issue for unsupervised topic modeling methods. As such, people have proposed evaluation metrics without relying on labels. We leverage the existing evaluation metrics, pointwise mutual information [Newman et al., 2010] and intrusion detection [Chang et al., 2009] that are proved to be effective in text-based topic modeling, and modify them to evaluate the multi-typed topic hierarchy. The metrics can be used to compare different methods in arbitrary datasets.

We evaluate the performance of the introduced methods on two datasets (see Table 2.2 for summary statistics of the constructed networks).

- **DBLP:** We take 33,313 recently published computer science papers from a subset of DBLP, which contain papers published in 20 conferences related to the areas of Artificial Intelligence, Databases, Data Mining, Information Retrieval, Machine Learning, and Natural Language Processing. We construct a heterogeneous network with three node types: term (from paper title), author and venue, and five link types: term-term, term-author, term-venue, author-author, and author-venue. As a paper is always published in exactly one venue, there can naturally be no venue-venue links.

- **NEWS:** We take 43,168 news articles on 16 top stories in 2013 from Google News. Text contents are extracted from html pages by heuristic rules, and an information extraction algorithm [Li et al., 2013] is used to extract entities. The 16 topics are: Bill Clinton, Boston Marathon, Earthquake, Egypt, Gaza, Iran, Israel, Joe Biden, Microsoft, Mitt Romney, Nuclear power, Steve Jobs, Sudan, Syria, Unemployment, US Crime. We construct a heterogeneous network with three node types: term (from article title), person and location, and six link types: term-term, term-person, term-location, person-person, person-location, and location-location.

The evaluation is twofold: (i) we evaluate the efficacy of subtopic discovery given a topic and its associated heterogeneous network; and (ii) we perform several "intruder detection" tasks to evaluate the quality of the constructed hierarchy based on human judgment.

2.3.1 EFFICACY OF SUBTOPIC DISCOVERY

We first present a set of experiments designed to evaluate just the subtopic discovery step (Step 2 in Section 2.2). It is the fundamental element of the recursive framework.

Evaluation measure: We extend the pointwise mutual information (PMI) metric in order to measure the quality of the multi-typed topics. The metric of pointwise mutual information PMI has been proposed by Newman et al. [2010] as a way of measuring the semantic coherence of topics. It is generally preferred over other quantitative metrics such as perplexity or the likelihood of held-out data. In order to measure the quality of the multi-typed topics, we extend the definition of PMI as follows.

For each topic, PMI calculates the average relatedness of each pair of the words ranked at top-K:

$$PMI(\mathbf{v}, \mathbf{v}) = \frac{2}{K(K-1)} \sum_{1 \le i < j \le K} \log \frac{p(v_i, v_j)}{p(v_i)p(v_j)}, \tag{2.44}$$

where $PMI \in [-\infty, \infty]$ and $\mathbf{v}$ are the top K most probable words of the topic. $PMI = 0$ implies that these words are independent; $PMI > 0 \, (< 0)$ implies they are overall positively (negatively) correlated.

However, the multi-typed topic contains not only words, but also other types of entities. So we define *heterogeneous* pointwise mutual information as:

$$HPMI(\mathbf{v}^{\mathbf{x}}, \mathbf{v}^{\mathbf{y}}) = \begin{cases} \frac{2}{K(K-1)} \sum_{1 \le i < j \le K} \log \frac{p(v_i^x, v_j^y)}{p(v_i^x)p(v_j^y)} & x = y \\ \frac{1}{K^2} \sum_{1 \le i, j \le K} \log \frac{p(v_i^x, v_j^y)}{p(v_i^x)p(v_j^y)} & x \ne y \end{cases}, \tag{2.45}$$

where $\mathbf{v}^{\mathbf{x}}$ are the top K most probable type-x nodes in the given topic. When $x = y$, HPMI reduces to PMI. The HPMI-score for every link type (x, y) is calculated and averaged to obtain an overall score. We set $K = 20$ for all node types.[1]

---

[1]The one exception is venues, as there are only 20 venues in the DBLP dataset, so we set $K = 3$ in this case.

Methods for comparison:

- **CATHYHIN (equal weight)** – The weight for every link type is set to be 1.

- **CATHYHIN (learn weight)** – The weight of each link type is learned, as described in Section 2.2.2. No parameters need hand tuning.

- **CATHYHIN (norm weight)** – The weight of each link type is explicitly set as: $\alpha_{x,y} = \frac{1}{\sum_{i,j} e_{i,j}^{x,y}}$. This is a heuristic normalization which forces the total weight of the links for each link type to be equal.

- **NetClus** – A state-of-the-art clustering and ranking method for heterogeneous networks. The link-type weight learning method for CATHYHIN does not apply to NetClus because NetClus does not have a single objective function to optimize.

- **TopK** – Select the top K nodes from each type according to their frequency to form a pseudo topic. This method serves as a baseline value for the proposed HPMI metric.

We discover the subtopics of four datasets:

- DBLP (20 conferences) – Aforementioned DBLP dataset. This dataset is used for evaluating the performance when constructing the first level of the hierarchy.

- DBLP (database area) – A subset of the DBLP dataset consisting only of papers published in five database conferences. By using this dataset, we analyze the quality of discovered subtopics in a lower level of the hierarchy.

- NEWS (16 topics) – Aforementioned NEWS dataset.

- NEWS (4 topic subset) – A subset of the NEWS dataset limited to 4 topics, which center around different types of entities: Bill Clinton, Boston Marathon, Earthquake, Egypt.

We show the heterogeneous pointwise mutual information averaged over the learned topics in Tables 2.3 and 2.4, our generative model consistently posts a higher HPMI score than Net-Clus (and TopK) across all links types in every dataset. Although NetClus HPMI values are better than the TopK baseline, the improvement of our best performing method—CATHYHIN (learn weight)—over the TopK baseline are better than the improvement posted by NetClus by factors ranging from 2–5.8. Even the improvement over the TopK baseline of CATHYHIN (equal weight), which considers uniform link type weights, is better than the improvement posted by NetClus by factors ranging from 1.6–4.6.

CATHYHIN with learned link-type weights consistently yields the highest overall HPMI scores, although CATHYHIN with normalized link type weights sometimes shows a slightly higher score for particular link types (e.g., Author-Author for both DBLP datasets, and Person-Person for both NEWS datasets). CATHYHIN (norm weight) assigns a high weight to a link

Table 2.3: Heterogeneous pointwise mutual information in DBLP (20 conferences and database area)

DBLP (Database Area)	Term-Term	Term-Author	Author-Author	Term-Venue	Author-Venue	Overall
Top K	-0.5228	-0.1069	0.4545	0.0348	-0.3650	-0.0761
NetClus	-0.3962	0.0479	0.4337	0.0368	-0.2857	0.0260
CATHYHIN (equal weight)	0.0561	0.4799	0.6496	0.0722	-0.0033	0.3994
CATHYHIN (norm weight)	-0.1514	0.3816	**0.6971**	0.0408	**0.2464**	0.3196
CATHYHIN (learn weight)	**0.3027**	**0.6435**	0.5574	**0.1165**	0.1805	**0.5205**

DBLP (20 Conferences)	Term-Term	Term-Author	Author-Author	Term-Venue	Author-Venue	Overall
Top K	-0.4825	-0.0204	0.5466	-1.0051	-0.4208	-0.0903
NetClus	-0.1995	0.5186	0.5404	0.2851	1.2659	0.4045
CATHYHIN (equal weight)	0.2936	0.8812	0.6595	0.5191	1.0466	0.6949
CATHYHIN (norm weight)	0.1825	0.8674	**0.9476**	0.7472	1.3307	0.7601
CATHYHIN (learn weight)	**0.4964**	**1.0618**	0.7161	**1.1283**	**1.7511**	**0.9168**

type whose total link weights were low in the originally constructed network, pushing the discovered subtopics to be more dependent on that link type. Normalizing the link-type weights does improve CATHYHIN performance in many cases, as compared to using uniform link-type weights. However, this heuristic determines the link type weight based solely on their link density. It can severely deteriorate the coherence of desne but valuable link types, such as Term-Term in both DBLP datasets, and rely too heavily on sparse but uninformative entities, such as Venues in the Database subtopic of the DBLP dataset.

Figure 2.8 demonstrates the learned link weights by CATHYHIN (learn weight) on DBLP datasets. At the first level, the term-venue and author-venue link types are assigned high weight, because the venue is a most important discriminator for general areas. At the second level, the venue links are much less useful for discovering subtopics in each area.

We may conclude from these experiments that CATHYHIN's unified generative model consistently outperforms the state-of-the-art heterogeneous network analysis technique NetClus. In order to generate coherent, multi-typed topics at each level of a topical hierarchy, it is important to learn the optimal weights of different entity types, which depends on the link-type density, the granularity of the topic to be partitioned, and the specific domain.

2.3.2 TOPICAL HIERARCHY QUALITY

The second set of evaluations assesses the ability of the method to construct a hierarchy of multi-typed topics that human judgement deems to be high quality. We generate and analyze multi-

Table 2.4: Heterogeneous pointwise mutual information in NEWS (16 topics collection and 4 topics subset)

NEWS (4 topics subset)	Term-Term	Term-Author	Person-Person	Term-Location	Person-Location	Location-Location	Overall
TopK	-0.2479	0.1671	0.0716	0.0787	0.2483	0.3632	0.1317
NetClus	0.1279	0.3835	0.2909	0.3240	0.4728	0.4271	0.3575
CATHYHIN (equal weight)	**1.0471**	0.7917	0.4902	0.8506	0.6821	0.6586	0.7610
CATHYHIN (norm weight)	0.7975	0.8825	**0.5553**	0.8682	**0.8077**	0.7346	0.8023
CATHYHIN (learn weight)	0.9935	**0.9354**	0.5142	**0.9784**	0.7389	**0.7645**	**0.8434**
NEWS (16 topics subset)	Term-Term	Term-Author	Person-Person	Term-Location	Person-Location	Location-Location	Overall
TopK	-1.7060	-0.8663	-0.8462	-1.0238	-0.5665	-0.4578	-0.8783
NetClus	-0.3847	0.0943	0.0313	-0.1114	0.1291	0.1376	-0.0274
CATHYHIN (equal weight)	0.7804	1.0170	0.8393	0.8354	0.9467	0.6382	0.8749
CATHYHIN (norm weight)	0.8579	**1.1143**	**0.9086**	0.8530	0.9624	**0.7143**	0.9284
CATHYHIN (learn weight)	**0.9234**	1.1109	0.7966	**0.9731**	**0.9718**	0.6965	**0.9500**

typed topical hierarchies using the DBLP dataset (20 conferences) and the NEWS dataset (16 topics collection).

We design three *Intruder Detection* tasks for human evaluation. Each task involves a set of questions asking humans to discover the "intruder" object from several options. Three annotators manually completed each task, and their evaluation scores were pooled.

The first task is Phrase Intrusion, which evaluates how well the hierarchies are able to separate phrases in different topics. Each question consists of X ($X = 5$ in our experiments) phrases; $X - 1$ of them are randomly chosen from the top phrases of the same topic and the remaining phrase is randomly chosen from a sibling topic. The second task is Entity Intrusion, a variation that evaluates how well the hierarchies are able to separate entities present in the dataset in different topics. For each entity type, each question consists of X entity patterns; $X - 1$ of them are randomly chosen from the top patterns of the same topic and the remaining entity pattern is randomly chosen from a sibling topic. This task is constructed for each entity type in each dataset (Author and Venue in DBLP; Person and Location in NEWS). The third task is Topic

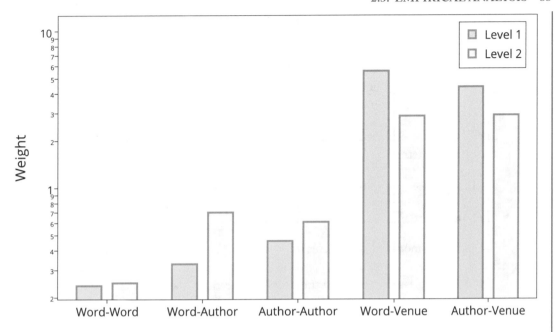

Figure 2.8: Learned link weights in DBLP.

Intrusion, which tests the quality of the parent-child relationships in the generated hierarchies. Each question consists of a parent topic t and X candidate child topics. $X - 1$ of the child topics are actual children of t in the generated hierarchy, and the remaining child topic is not. Each topic is represented by its top 5 ranked patterns of each type—e.g., for the NEWS dataset, the top 5 phrases, people, and locations are shown for each topic.

For DBLP, we generate 210 phrase intrusion questions, 210 entity instrusion questions for both author and venue type, and 60 topic intrusion questions. For NEWS, we generate 280 phrase intrusion questions, 280 entity instruction questions for both person and location type, and 100 topic intrusion questions. Figure 2.9 shows examples of generated questions in DBLP. For each question, 3 human annotators with background knowledge of computer science and news select the intruder phrase, entity, or subtopic. If they are unable to make a choice, or choose incorrectly or inconsistently, the question is marked as a failure.

Methods for comparison:

- **CATHYHIN** – As defined in Section 2.2.

- **CATHYHIN$_1$** – The pattern length of text and every entity type is restricted to 1

- **CATHY** – As defined in Section 2.1, the hierarchy is constructed only from textual information.

- **CATHY$_1$** – The phrase length is restricted to 1.

- **CATHYheuristic_HIN** – Since neither CATHY nor CATHY$_1$ provides topical ranks for entities, we construct this method to have a comparison for the Entity Intrusion task. We use a heuristic entity ranking method based on the textual hierarchy generated by CATHY, and the original links in the network (see Chapter 4).

- **NetClusphrase** – NetClus is used for subtopic discovery, followed by the topical mining and ranking method of CATHYHIN, as described in Chapter 3.

- **NetClusphrase$_1$** – Equivalent to NetClusphrase with the phrase length restricted to 1.

- **NetClus** – As defined in Sun et al. [2009].

The optimal smoothing parameter for NetClus is $\lambda_S = 0.3$ and 0.7 in DBLP and NEWS, respectively.

Table 2.5 displays the results of the intruder detection tasks. For the Entity Intrusion task on the DBLP dataset, we restricted the entity pattern length to 1 in order to generate mean-

Phrase Intrusion			
Question 1/210 data mining	association rules	logic programs	data streams
Question 2/210 natural language	query optimization	data management	database sstems

Venue Intrusion			
Question 1/210 KDD	SDM	ICDE	ICDM
Question 2/210 EMNLP	ACL	AAAI	HLT-NAACL

Question 1/60	Topic Intrusion			
Parent topic	**Child topic 1**	**Child topic 2**	**Child topic 3**	**Child topic 3**
database systems	web search	data management	query processing	database system
data management	search engine	data integration	query optimization	database design
query processing	semantic web	data sources	query databases	expert system
management system	search results	data warehousing	relational databases	management system
data system	web pages	data applications	query data	design system

Figure 2.9: Examples of Intruder Detection questions.

ingful questions. This renders the methods CATHYHIN$_1$ and NetClusphrase$_{-1}$ equivalent to CATHYHIN and NetClusphrase, respectively, so we omit the former methods from reporting.

Table 2.5: Results of Intruder Detection tasks (% correct intruders identified)

	DBLP				NEWS			
	Phrase	Venue	Author	Topic	Phrase	Location	Person	Topic
CATHYHIN	**0.83**	**0.83**	**1.0**	**1.0**	**0.65**	**0.70**	**0.80**	**0.90**
CATHYHIN$_1$	0.64	–	–	0.92	0.40	0.55	0.50	0.70
CATHY	0.72	–	–	0.92	0.58	–	–	0.65
CATHY$_1$	0.61	–	–	0.92	0.23	–	–	0.50
CATHY$_{heur\_HIN}$	–	0.78	0.94	0.92	–	0.65	0.45	0.70
NetClus$_{pattern}$	0.33	0.78	0.89	0.58	0.23	0.20	0.55	0.45
NetClus$_{pattern\_1}$	0.53	–	–	0.58	0.20	0.45	0.30	0.40
NetClus	0.19	0.78	0.83	0.83	0.15	0.35	0.25	0.45

Experiment results: The Phrase Intrusion task performs much better when phrases are used rather than unigrams, for both CATHYHIN and CATHY, on both datasets. The NEWS dataset exhibits a stronger preference for phrases, as opposed to the DBLP dataset, which may be due to the fact that the terms in the NEWS dataset are more likely to be noisy and uninformative outside of their context, whereas the DBLP terms are more technical and therefore easier to interpret. This characteristic may also help explain why the performance of every method on DBLP data is consistently higher than on NEWS data. However, neither phrase mining and ranking nor unigram ranking can make up for poor performance during the topic discovery step, as seen in the three NetClus variations. Therefore, both phrase representation and high quality topics are necessary for good topic interpretability.

For the Entity Intrusion task, all of the relevant methods show comparable performance in identifying Author and Venue intruders in the DBLP dataset (although CATHYHIN is still consistently the highest). Since the DBLP dataset is well structured, and the entity links are highly trustworthy, identifying entities by topic is likely easier. However, the entities in the NEWS dataset were automatically discovered from the data, and the link data is therefore noisy and imperfect. CATHYHIN is the most effective in identifying both Location and Person intruders. Once again, both better topic discovery and improved pattern representations are responsible for CATHYHIN's good results, and simply enhancing the pattern representations, whether for CATHY or NetClus, cannot achieve competitive performance.

CATHYHIN performs very well in the Topic Intrusion task on both datasets. Similar to the Phrase Intrusion task, both CATHYHIN and CATHY yield equally good or better result when phrases and entity patterns are mined, rather than just terms and single entities. The fact that CATHYHIN always outperforms CATHY demonstrates that utilizing entity link information is indeed helpful for improving topical hierarchy quality. In all three intruder detection tasks on both datasets, CATHYHIN consistently outperforms all other methods, showing that an integrated heterogeneous model consistently produces a more robust hierarchy which is more easily interpreted by human judgment.

2.3.3 CASE STUDY

Table 2.6: The "information retrieval" topic, as generated by three methods

CATHYHIN	CATHY$_{heuristic\_HIN}$	Net Clus$_{pattern}$
{information retrieval; web search; retrieval} / {W. Bruce Croft; Iadh Ounis; James Allen} / {SIGIR; WWW; ECIR}	{information retrieval; search engine; web search} / {Ryen W. White; C. Lee Giles; Mounia Lalmas} / {SIGIR; WWW; ECIR}	{information retrieval; statistical machine translation; conditional random fields} / {W. Bruce Croft; Zheng Chen; Chengxiang Zhai} / {ACL; SIGIR; HLT-NAACL}

In Section 2.3.2, we analyzed the two main reasons for the performance gain of CATHY-HIN: improved pattern representations and better topic discovery. In Figure 2.5 we showed real examples of the constructed hierarchy in the DBLP data. It is clear that the multi-typed entities enrich the context of each topic and improve the representations of text-only topics. Here we use one simple example to illustrate the topic discovery performance, using the same topic representations.

Table 2.6 illustrates three representations of the topic "information retrieval" (one of the six areas in DBLP dataset). Overall, CATHYHIN finds more "pure" information retrieval entities. CATHYheuristic_HIN generates very similar top-ranked phrases and venues, but different authors. While the top authors found by CATHYheuristic_HIN indeed work on information retrieval, they spread their interest in other fields too, such as data mining and human computer interaction. This is because CATHYheuristic_HIN only uses the links between text and entities to rank entities posterior to the text-based topic discovery, while CATHYHIN can further use the author-author and author-venue links to refine the topics and find the more accurate position for each entity. NetCluspattern also utilizes the multiple types of links, but it mixes different topics, such as information retrieval and natural language processing, perhaps due to the hard partitioning of papers and heuristic combination of ranking and clustering.

As a worst-case study, Table 2.7 illustrates three representations of the topic "Egypt" (one of the 16 top stories in NEWS dataset), each with its least comprehensible subtopic. The locations found within the CATHYHIN subtopic are sensible. However, CATHYheuristic_HIN first constructs phrase-represented topics from text, and then uses entity link information to rank entities in each topic. Thus, the entities are not assured to fit well into the constructed topic, and indeed, the CATHYheuristic_HIN subtopic's locations are not reasonable given the parent topic. For example, CATHY discovers *"supreme leader/army general sex/court/supreme court/egypts prosecutor general"* to be a subtopic of *"egypt/egypts morsi/egypt imf loan/egypts president/muslim brotherhood."* Resorting to the original network links to discover each topic's entity rankings results in the claim that the locations *"US/Sudan/Iran/Washington"* represent a subtopic of locations *"Egypt/Cairo/Tahrir Square/Port Said,"* which is not easily interpretable. Finally, NetClus_pattern conflates "Egypt" with several other topics, and the pattern representations can do little to improve the topic interpretability.

Table 2.7: The "Egypt" topic and the least sensible subtopic, as generated by three methods (only Phrases and Locations are shown)

CATHYHIN	**CATHY**_heuristic_HIN_	**Net Clus**_pattern_
{egypt; egypts; death toll; morsi} / {Egypt; Egypt Cairo; Egypt Israel; Egypt Gaza}	{egypt; egypts morsi; egypt imf loan; egypts president} / {Egypt; Cairo; Tahrir Square; Port Said}	{bill clinton; power nuclear; rate unemployment; south sudan} / {Egypt Cairo; Egypt Coptic; Israel Jerusalem; Libya Egypt}
↓	↓	↓
{death toll; egyptian; sexual harassment; egypt soccer} / {Egypt Cairo; Egypt Gaza; Egypt Israel}	{supreme leader; army general sex; court; supreme court} / {US; Sudan; Iran; Washington}	{egypts coptic pope; egypts christians; obama romney; romney campaign} / {Egypt Cairo; Egypt Coptic; Israel Jerusalem; Egypt}

CHAPTER 3

Topical Phrase Mining

A topic is traditionally modeled as a multinomial distribution over terms, and frequent terms related by a common theme are expected to have a large probability in a topic multinomial.

When latent topic multinomials are inferred, it is of interest to visualize these topics in order to facilitate human interpretation and exploration of the large amounts of unorganized text often found within text corpora. In addition, visualization provides a qualitative method of validating the inferred topic model [Chang et al., 2009].

A good visualization requires mining phrases that are both high quality and topic-representative.

Generally speaking, mining quality phrases refers to automatically extracting salient phrases from a given corpus. It is a fundamental task for text analytics of various domains, such as science, news, social media, and enterprise documents. In these large, dynamic collections of documents, analysts are often interested in variable-length phrases, including scientific concepts, events, organizations, products, slogans, and so on. Efficient extraction of quality phrases enable a large body of applications to transform from word granularity to phrase granularity. Examples of such applications include topic tracking, OLAP on multi-dimensional text collections, and document clustering. For keyword search, the extracted phrases facilitate selective indexing, query suggestion, and other tasks. Also, extraction of phrases is critical towards information extraction because many concepts, entities, and relations are manifested in phrases. Although the study of this task originates from the natural language processing (NLP) community, the challenge has been recognized of applying NLP tools in the emerging big data that deviate from rigorous language rules. Query logs, social media messages, tags, and textual transaction records are just a few examples.

Furthermore, the quality phrases need to be further selected to represent mined topics. While it is appealing to incorporate the phrase-finding element in the topical clustering process, these methods often suffer from high-complexity, and overall demonstrate poor scalability outside small datasets.

In this chapter, new methods are introduced that demonstrate both scalability compared to other topical phrase mining methods and interpretability. Frequent pattern mining and statistical analysis are employed for phrase mining. We leverage the redundancy of the text in the corpus, rather than relying on linguistic analysis. Therefore, these methods do not require domain knowledge or language grammar. Also, these methods can work with informally written text such as social media messages.

We introduce some notations first. We use $C[\cdot]$ to index a word in the corpus string and $|C|$ to denote the corpus size. For convenience we index all the unique words in this corpus using a vocabulary of V words. And $C[i] = x, x \in \{1, \ldots, V\}$ means that the i-th token in the corpus is the x-th word in the vocabulary. Throughout this chapter we use "word x" to refer to the x-th word in the vocabulary.

A phrase is defined to be a sequence of words $P = \{v_i\}_{i=1}^n$. The length n can be any positive integer.

3.1 CRITERIA OF GOOD PHRASES AND TOPICAL PHRASES

Our goal is to develop a mining algorithm for finding high quality candidate phrases, and design a ranking function to evaluate the quality of topical phrases. To do so, we need to understand human intuition for judging what constitutes a high quality topical phrase.

First, we notice that it is inappropriate to discard all unigrams when approaching this task, or in fact to demonstrate a bias towards any particular phrase length. For instance, consider that the unigram "classification" and the trigram "support vector machines" are both high quality topical phrases for the machine learning topic in the domain of computer science. It is also not ideal to present separate ranked lists of topical phrases of each length, since when people are asked to characterize topics, they do not limit themselves to, e.g., listing only bigrams, but rather provide a set of relevant phrases with no regard for phrase length. We should therefore also be able to directly compare phrases of mixed-length in a natural way. We refer to this characteristic as exhibiting the **comparability property**.

Traditional probabilistic modeling approaches, such as language models or topic models do not have the comparability property. They can model the probability of seeing an n-gram given a topic, but the probabilities of n-grams with different lengths (unigrams, bigrams, etc.) are not well comparable. These approaches simply find longer n-grams to have much smaller probability than shorter ones, because the probabilities of seeing every possible unigram sum up to 1, and so do the probabilities of seeing every possible bigram, trigram, etc. However, the total number of possible n-grams grows following a power law ($\mathcal{O}(V^n)$), and ranking functions based on these traditional approaches invariably favor short n-grams. While previous work has used various heuristics to correct this bias during post-processing steps, such as using a penalization term with respect to the phrase length [Tomokiyo and Hurst, 2003; Zhao et al., 2011], we will introduce a more principled approach.

The key to exhibiting the comparability property is to treat each phrase as a whole unit, and design quality measure for them.

We specify two requirements of a good phrase.

- **Concordance:** In corpus linguistics, concordance refers to the co-occurrence of tokens in such frequency that is significantly higher than what is expected due to chance. A commonly used example of a phraseological-concordance is the example of the two candidate phrases "strong tea" and "powerful tea" [Halliday et al., 1966]. One would assume that the two phrases appear in similar frequency, yet in the English language, the phrase "strong tea" is considered more proper and appears in much higher frequency. Because a concordant phrase's frequency deviates from what is expected, we consider them "interesting" and informative. This insight motivates the necessity of analyzing phrases probabilistically to ensure they are concordant.

- **Completeness:** If a long phrase and its subsets may both satisfy the above criterion. For example, in the case of "mining frequent patterns," "mining frequent" will satisfy the concordance restriction, yet is clearly a subset of a larger and more intuitive phrase. Our phrase-construction algorithm should be able to automatically determine the most appropriate size for a human-interpretable phrase.

Besides the general requirements of phrases, we wish to find high quality topical phrases that successfully represent a topic. We now present the characteristics that such a phrase should have. As a running example, consider mining and ranking phrases for topics in Computer Science.

- **Popularity:** Popularity, which may be referred to by other names such as coverage, frequency, or importance, is the most basic criteria, required by every ranking function that tackles this same problem. *Example: "information retrieval" has better coverage than "cross-language information retrieval" in the Information Retrieval topic.* A phrase that is not frequent in a topic should never be highly ranked as being representative of that topic, regardless of its length, or its value according to any other criteria. This further suggests that the ranking function should be designed in such a way that a topical phrase with low popularity is guaranteed to have a lower rank.

- **Purity:** A phrase is pure in a topic if it is only frequent in documents belonging to that topic and not frequent in documents within other topics. *Example: "query processing" is more pure than "query" in the Database topic.* Like popularity, some version of this criterion is also present in most ranking functions, though it might be referred by other names, such as "informativeness" [Tomokiyo and Hurst, 2003] or "relevance" [Zhao et al., 2011].

A phrase mining algorithm needs to carry out these criteria when finding high quality phrases. We present two specific mining algorithms in the next two sections.

3.2   KERT: MINING PHRASES IN SHORT, CONTENT-REPRESENTATIVE TEXT

This section focuses on mining topical phrases from a collection of short documents. There are many cases where the full text of a document collection is not available, or is too noisy, for the desired task of topic discovery from the collection. Furthermore, we focus on documents which are information-rich, meaning that most of the document content is informative, not noisy, with the usual exception of function words. The documents may also be a mix of multiple topics, in spite of their short length. Scientific paper titles fit this criterion well. While the framework could technically be applied to a collection of noisy short documents such as tweets, it would require at least a transformation of the noisy documents into information-rich documents in order to perform well.

3.2.1   PHRASE QUALITY

Concordance

A group of words should be grouped into a phrase if they co-occur significantly more frequently than the expected co-occurrence frequency given that each word in the phrase occurs independently. For example, while "active learning" is a good phrase in the Machine Learning topics, "learning classification" is not, since the latter two words co-occur only because both of them are popular in the topic.

We therefore compare the probability of seeing a phrase $P = \{v_1 \dots v_n\}$ and seeing the n words $v_1 \dots v_n$ independently:

$$
\begin{aligned}
\kappa^{con}(P) &= \log \frac{p(P)}{\prod_{v \in P} p(v)} \\
&= \log \frac{f(P)}{N} - \sum_{v \in P} \log \frac{f(v)}{N},
\end{aligned}
\tag{3.1}
$$

where N is the total number of documents.

Completeness

A phrase P is not complete if a longer phrase P' that contains P usually co-occurs with P. For example, while "support vector machines" is a complete phrase, "vector machines" is not, as "support" nearly always accompanies "vector machines."

We thus measure the completeness of a phrase P by examining the conditional probability of observing P' given P in a topic-t document:

$$
\begin{aligned}
\kappa^{com}(P) &= 1 - \max_{P' \supsetneq P} p(P'|P) \\
&= 1 - \max_{v} p(P \oplus v|P) \\
&= 1 - \frac{\max_v f(P \oplus v)}{f(P)}.
\end{aligned}
\tag{3.2}
$$

3.2.2   TOPICAL PHRASE QUALITY

We first introduce a notion of *topical frequency* of phrases.

Definition 3.1   Topical Frequency.   The topical frequency $f_t(P)$ of a phrase is the count of the times the phrase is attributed to topic t. For the root node o, $f_o(P) = f(P)$ is equal to the total frequency. For each topic node in the hierarchy, with C_t subtopics, $f_t(P) = \sum_{z \in [C_t]} f_{t \odot z}(P)$, i.e., the topical frequency is equal to the sum of the sub-topical frequencies.

Table 3.1:   Example of topical frequency estimation. The topics are assumed to be inferred as machine learning, database, data mining, and information retrieval from the collection

Phrase	ML	DB	DM	IR	Total
support vector machines	85	0	0	0	85
query processing	0	212	27	12	251
world wide web	0	7	1	26	34
social networks	39	1	31	33	104

Table 3.1 illustrates an example of estimating topical frequency for phrases in a computer science topic that has four subtopics. The phrase "support vector machines" is estimated to belong entirely to the Machine Learning (ML) topic with high frequency, and therefore is a candidate for a high quality phrase. However, "social networks" is fairly evenly distributed among three topics, and is thus less likely to be a high quality phrase.

Using the learned topic model parameters in the last section, we estimate the topical frequency of a phrase $P = \{v_1 \dots v_n\}$ based on two assumptions: (i) for a topic-t phrase of length n, each of the n words is generated with the distribution ϕ^t and (ii) the total number of topic-t phrases of length n is proportional to $\rho_{\pi_t, t}$:

$$f_{t \odot z}(P) = f_t(P) \frac{\rho_{t \odot z} \prod_{i=1}^{n} \phi_{v_i}^{t \odot z}}{\sum_{c \in C_t} \rho_{t \odot c} \prod_{i=1}^{n} \phi_{v_i}^{t \odot c}}. \tag{3.3}$$

We illustrate how to use the topical frequency to quantify the two criteria about topical representativeness in Section 3.1 and combine them.

Popularity

A representative phrase for a topic should cover many documents within that topic. For example, "information retrieval" has better coverage than "cross-language information retrieval" in the topic of Information Retrieval. We directly quantify the popularity measure of a phrase as the probability of seeing a phrase in a random topic-t document $p(P|t)$:

$$\kappa_t^{pop}(P) = p(P|t) = \frac{f_t(p)}{N_t},$$
(3.4)

where N_t be the number of documents that contain at least one frequent topic-t phrase with topical frequency larger than a threshold μ.

Purity

A phrase is pure in topic t if it is only frequent in documents about topic t and not frequent in documents about other topics. For example, "query processing" is a more pure phrase than "query" in the Databases topic.

We measure the purity of a phrase by comparing the probability of seeing a phrase in the topic-t documents and the probability of seeing it in any other topic-t' collection ($t' = 0, 1, \ldots, k$, $t' \neq t$). If there exists a topic t' such that the probability of $p(P|t')$ is similar or even larger than $p(P|t)$, the phrase P indicates confusion about topic t and t'. The purity of a phrase compares the probability of seeing it in the topic-t collection and the maximal probability of seeing it in any mix collection:

$$\kappa_t^{pur}(P) = \log \frac{p(P|t)}{\max_{t'} p(P|\{t, t'\})}$$
$$= \log \frac{f_t(P)}{N_t} - \log \max_{t'} \frac{f_t(P) + f_{t'}(P)}{N_{\{t,t'\}}},$$
(3.5)

where $N_{\{t,t'\}}$ is the number of documents that contain at least one frequent topic-t phrase or topic-t' phrase.

We can combine these two measures into a function representing the quality of a topical phrase by viewing them within an information theoretic framework. As described above, the popularity criterion is in some sense the more important, since a phrase with low popularity will be necessarily of low quality, regardless of its performance according to the other criteria. We can enforce this by representing the relationships between popularity and purity using a pointwise Kullback-Leibler (KL)-divergence metric.

Pointwise KL-divergence is a distance measure between two probabilities that takes the absolute probability into consideration, and is more robust than pointwise mutual information when the relative difference between probabilities need to be supported by sufficiently high absolute probability.

The product of popularity and purity, $\kappa_t^{pop}(P)\kappa_t^{pur}(P) = p(P|t) \log \frac{p(P|t)}{p(P|\{t,t^*\})}$ is equal to the pointwise KL-divergence between the probabilities of $p(P|t)$ and $p(P|\{t, t^*\})$.

Likewise, the product of popularity and concordance, $\kappa_t^{pop}(P)\kappa_t^{con}(P)$ is equivalent to the pointwise KL-divergence between the probability of $p(P|t))$ under different independence assumptions.

Finally, we combine the two pointwise KL-divergence with a weighted summation, and implement the completeness criterion as a filtering condition to remove incomplete phrases:

$$Quality_t(P) = \begin{cases} 0 & \kappa_t^{com} \leq \gamma \\ \kappa_t^{pop}[(1-\omega)\kappa_t^{pur} + \omega\kappa^{con}](P) & \text{o.w.} \end{cases} . \tag{3.6}$$

Here, $\gamma \in [0, 1]$ controls how aggressively we prune incomplete phrases. $\gamma = 0$ corresponds to ignoring the completeness criterion and retaining all *closed* phrases, where no supersets have the same topical support. As γ approaches 1, more phrases will be filtered and eventually only *maximal* phrases (no supersets are sufficiently frequent) will remain. The other three criteria rank phrases that pass the completeness filter.

Although $Quality_t(P)$ is a combination of two pointwise KL-divergence metrics, the coverage criterion is a factor in both. This reflects the fact that when $p(P|t)$ is small, phrase P has low support, and thus the estimates of purity and concordance will be unreliable and their role should be limited. When $p(P|t)$ is large, phrase P has high support, and magnifies the cumulative effect (positive or negative) of the purity and concordance criteria.

The relative weights of the purity and concordance criteria are controlled by $\omega \in [0, 1]$. Both measures are log ratios on comparable scales, and can thus be balanced by a weighted summation. As ω increases, we expect more topic-independent, but common phrases to be ranked higher.

3.3 TOPMINE: MINING PHRASES IN GENERAL TEXT

KERT is simple, yet relying on the assumption that the text is short and information rich. In this section, we propose a new methodology ToPMine that works well with generic text.

KERT filters and ranks phrases according to frequency-based statistics. However, the raw frequency from the data may produce misleading quality assessment, as the following example demonstrates.

Example 3.2 Raw Frequency-based Phrase Mining Consider a set of scientific publications and the raw frequency counts of two phrases "relational database system" and "support vector machine" and their subsequences in the *frequency* column of Table 3.2. The numbers are hypothetical but emulate several key observations: (i) the frequency generally decreases with the phrase length; (ii) both good and bad phrases can possess high frequency (e.g., "support vector" and "vector machine"); and (iii) the frequency of one sequence (e.g., "relational database system") and its subsequences can have a similar scale of another sequence (e.g., "support vector machine") and its counterparts. ∎

Obviously, a method that solely ranks the phrases according to the frequency will output incomplete phrases such as "vector machine." In order to address this problem, KERT has proposed a heuristic based on comparison of a sequence's frequency and its sub and super sequences, assuming that a good phrase should have high enough (normalized) frequency compared with its constituent words and super phrases. However, such a heuristic can hardly differentiate the quality of, e.g., "support vector" and "vector machine" because their frequency are so close. Finally, even if the heuristic can indeed draw a line between "support vector" and "vector machine" by discriminating their frequency (between 160 and 150), the same separation could fail for another case like "relational database" and "database system."

Table 3.2: A hypothetical example of word sequence raw frequency

sequence	frequency	phrase?	rectified
relational database system	100	yes	70
relational database	150	yes	40
database system	160	yes	35
relational	500	N/A	20
database	1000	N/A	200
system	10000	N/A	1000
support vector machine	100	yes	80
support vector	160	yes	50
vector machine	150	no	6
support	500	N/A	150
vector	1000	N/A	200
machine	1000	N/A	150

Using the frequency in Table 3.2, all heuristics will produce identical predictions for "relational database" and "vector machine," guaranteeing one of them wrong. This example suggests the intrinsic limitations of using raw frequency counts, especially in judging whether a sequence is too long (longer than a minimum semantic unit), too short (broken and not informative), or right in length. It is a critical bottleneck for all frequency-based quality assessment.

ToPMine addresses this bottleneck, proposing to rectify the decisive raw frequency that hinders discovering the true quality of a phrase. The goal of the *rectification* is to estimate how many times each word sequence should be interpreted in whole as a phrase in its occurrence context. The following example illustrates this idea.

Example 3.3 Rectification   Consider the following occurrences of the six multi-word sequences listed in Table 3.2.

1. A ⌈relational database system⌋ for images...

2. ⌈Database system⌋ empowers everyone in your organization...

3. More formally, a ⌈support vector machine⌋ constructs a hyperplane...

4. The ⌈support vector⌋ method is a new general method of ⌈function estimation⌋...

5. A standard ⌈feature vector⌋ ⌈machine learning⌋ setup is used to describe...

6. ⌈Relevance vector machine⌋ has an identical ⌈functional form⌋ to the ⌈support vector machine⌋...

7. The basic goal for ⌈object-oriented relational database⌋ is to bridge the gap between...   ∎

The first four instances should provide positive counts to these sequences, while the last three instances should not provide positive counts to "vector machine" or "relational database" because they should not be interpreted as a whole phrase (instead, sequences like "feature vector" and "relevance vector machine" can). Suppose one can correctly identify whether to count for each occurrence of the sequences, and collect rectified frequency as shown in the *rectified* column of Table 3.2. The rectified frequency now clearly distinguishes "vector machine" from the other phrases, since "vector machine" rarely occurs independently as a phrase.

The success of this approach relies on reasonably accurate rectification. Simple arithmetics of the raw frequency, such as subtracting one sequence's count with its quality super sequence, are prone to error. One reason is that it is context dependent to decide whether a sequence should be considered as a phrase candidate, e.g., the fifth instance in Example 3.3 prefers "feature vector" and "machine learning" over "vector machine." Similar situation is more common for web data. Consider a query log like "⌈bank of america⌋ online." It is obvious that for the super sequences of "america online" (i.e., "of america online" and "bank of america online"), neither of them could be a quality phrase. Thus, the raw frequency of "america online" is hard to be rectified by simple subtraction.

The context information is lost when we only collect the frequency counts. In order to recover the true frequency with best effort, we ought to examine the context of every word sequence's occurrence and decide whether to count it as a phrase. The examination for one occurrence may involve enumeration of alternative possibilities, such as extending the sequence or breaking the sequence, and comparison among them. The test for word sequence occurrences could be expensive, losing the advantage in efficiency of the frequent pattern mining approaches.

Facing the challenge of accuracy and efficiency, we propose a segmentation approach named *phrasal segmentation*, and integrate it with the phrase quality assessment in a unified framework with linear complexity (w.r.t the corpus size). First, the segmentation assigns every word occurrence to only one phrase. In the first instance of Example 3.3, "relational database system" are bundled as a single phrase. Therefore, it automatically avoids double counting "relational database" and "database system" within this instance. Similarly, the segmentation of the fifth instance contributes to the count of "feature vector" and "machine learning" instead of "feature," "vector machine," and "learning." This strategy condenses the individual tests for each word sequence and reduces the overall complexity while ensures correctness. Second, though there are an exponential number of possible partitions of the documents, we are concerned with those relevant to the phrase extraction task only. Therefore, we can integrate the segmentation with the phrase quality assessment, such that (i) only frequent phrases with reasonable quality are taken into consideration when enumerating partitions; and (ii) the phrase quality guides the segmentation, and the segmentation rectifies the phrase quality estimation. Such an integrated framework leverages mutual enhancement, and achieves both high quality and high efficiency. Finally, both the phrase quality and the segmentation results are useful from an application point of view. The segmentation results are especially desirable for tasks like document indexing, clustering or retrieval.

To extract topical phrases that satisfy our desired requirements, we introduce the TopMind solution that can be divided into three main parts: phrase mining, text segmentation, and topical phrase ranking. The phrase mining part finds frequent patterns; text segmentation helps finding high-quality phrases using the two requirements of good phrases, and feed them to topical phrase ranking.

3.3.1   FREQUENT PHRASE MINING

In Algorithm 1, we present the frequent phrase mining algorithm. The task of frequent phrase mining can be defined as collecting aggregate counts for all contiguous words in a corpus that satisfy a certain minimum support threshold. We draw upon a property for efficiently mining these frequent phrases.

Prefix property: If phrase P is frequent, any prefix of P should be frequent too. In this way, all the frequent phrases can be generated by expanding their prefixes.

We can exploit this property for our case of phrases by maintaining a set of active indices. These active indices are a list of positions in the corpus at which a contiguous pattern of length n is frequent. In line 1 of Algorithm 1, we use *index* to maintain the active indices.

The frequency criterion requires phrases to have sufficient occurrences. In general, we can set a minimum support that grows linearly with corpus size. The larger minimum support is, the higher precision and the lower recall are expected.

Since the operation of Hash table is $O(1)$, both the time and space complexities are $O(L|\mathcal{C}|)$, where L is the maximum length of frequent phrases and can be regarded as a constant in most cases. This algorithm is linear to the size of corpus $|\mathcal{C}|$.

3.3.2   SEGMENTATION AND PHRASE FILTERING

The discussion in Example 3.2 points out the limitations of using only raw frequency counts. Instead, we ought to examine the context of every word sequence's occurrence and decide whether to count it as a phrase like in Example 3.3. Here we introduce an efficient phrasal segmentation method to compute rectified frequency of each phrase.

The key element of the method is a bottom-up merging process. At each iteration, the algorithm makes locally optimal decisions in merging single- and multi-word phrases as guided by a statistical significance score. In the following, we present an agglomerative phrase-construction algorithm, and then explain how the significance of a potential merging is evaluated and how this significance guides our agglomerative merging algorithm.

Algorithm 1: Frequent Phrase Detection

1.1 **Input**: Corpus $\mathcal{C}$, minimum support threshold μ.

1.2 **Output**: Raw frequency counter f of frequent phrases

1.3 $index \leftarrow$ an empty dictionary

1.4 $f \leftarrow$ an empty dictionary

1.5 **for** $i \leftarrow 1$ **to** $|\mathcal{C}|$ **do**

1.6 $index[\mathcal{C}[i]] \leftarrow index[\mathcal{C}[i]] \cup \{i\}$

1.7 **while** $index$ *is not empty* **do**

1.8 $index' \leftarrow$ an empty dictionary

1.9   **for** $P \in index.keys$ **do**

1.10   **if** $|index[P]| \geq \mu$ **then**

1.11 $f[P] \leftarrow |index[P]|$

1.12   **for** $p \in index[P]$ **do**

1.13 $P' \leftarrow P \oplus \mathcal{C}[p+1]$

1.14 $index'[P'] \leftarrow index'[P'] \cup \{p+1\}$

1.15 $index \leftarrow index'$

1.16 **return** f

We construct quality phrases by inducing a partition upon each document. We employ a bottom-up agglomerative merging that greedily merges the best possible pair of candidate phrases at each iteration. This merging constructs phrases from single- and multi-word phrases while maintaining the partition requirement. Because only phrases induced by the partition are valid phrases, we have implicitly filtered out phrases that may have passed the minimum support criterion by random chance.

In Algorithm 2, we present the phrase construction algorithm. The algorithm takes as input a document and the aggregate counts obtained from the frequent phrase mining algorithm. It then iteratively merges phrase instances with the strongest association as guided by a potential merging's significance measure. The process is a bottom-up approach that induces a partition upon the original document creating a "bag-of-phrases."

Figure 3.1 tracks the phrase construction algorithm by visualizing the agglomerative merging of phrases at each iteration with a dendogram. Operating on a paper title obtained from the DBLP titles dataset, each level of the dendogram represents a single merging. At each iteration, the algorithm selects two contiguous phrases such that their merging is of highest significance (Line 2.4) and merges them (Lines 2.6–2.8). The following iteration then considers the newly merged phrase as a single unit. By considering each newly merged phrase as a single unit and assessing the significance of merging two phrases at each iteration, it addresses the "free-rider"

Algorithm 2: Bottom-up Construction of Phrases from Ordered Tokens

---

Input: Raw frequency f, threshold α of merging
Output: Partition

2.1 $H \leftarrow MaxHeap()$
2.2   Place all contiguous token pairs into H with their significance score key
2.3   **while** $H.size() > 1$ **do**
2.4 $Best \leftarrow H.getMax()$
2.5   **if** $Best.Sig \geq \alpha$ **then**
2.6 $New \leftarrow Merge(Best)$
2.7 $H.Remove(Best)$
2.8   Update significance for New with its left phrase instance and right phrase instance
2.9   **else**
2.10 $break$

---

problem where long, unintelligible, phrases are evaluated as significant when comparing the occurrence of a phrase to the occurrence of each constituent word independently.

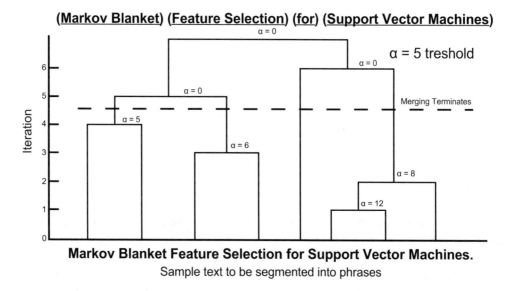

Figure 3.1: Bottom-up construction of a "bag of phrases."

The algorithm terminates when the next merging with the highest significance does not meet a predetermined significance threshold α or when all the terms have been merged into a single phrase. This is represented by the dashed line in Figure 3.1 where there are no more candidate phrases that meet the significance threshold. Upon termination, a natural "bag-of-phrases" partition remains. The phrase construction algorithm addresses both concordance and completeness criteria.

To statistically reason about the occurrence of phrases, we consider a null hypothesis, that the corpus is generated from a series of independent Bernoulli trials. Under this hypothesis, the presence or absence of a phrase at a specific position in the corpus is a product of a Bernoulli random variable. It follows that the expected number of occurrences of a phrase can be interpreted as a binomial random variable. Because the number of tokens $L = |\mathcal{C}|$ in the corpus can be assumed to be fairly large, the binomial can be reasonably approximated by a normal distribution. The count $f(P)$ of a phrase P within the corpus is a random variable following the null hypothesis distribution:

$$f(P) \sim \mathcal{N}(Lp(P), Lp(P)(1 - p(P)))) \approx \mathcal{N}(Lp(P), Lp(P)),$$

where $p(P)$ is the Bernoulli trial success probability for phrase P. The empirical probability of a phrase in the corpus can be estimated as $p(P) \approx \frac{f(P)}{L}$.

Consider a longer phrase that comprises of two phrases P_1 and P_2. The mean of its frequency under our null hypothesis of independence of the two phrases as:

$$\mu_0(f(P_1 \oplus P_2)) = Lp(P_1)p(P_2).$$

This expectation follows from treating each phrase as a constituent, functioning as a single unit in the syntax. Due to the unknown population variance and sample-size guarantees from the minimum support, we can estimate the variance of the population using sample variance: $\sigma_{P_1 \oplus P_2}^2 \approx f(P_1 \oplus P_2)$, the sample phrase occurrence count.

We use a significance score to provide a quantitative measure of which two consecutive phrases form the best collocation at each merging iteration. This is measured by comparing the actual frequency with the expected occurrences under the null hypothesis:

$$sig(P_1, P_2) \approx \frac{f(P_1 \oplus P_2) - \mu_0(P_1, P_2)}{\sqrt{f(P_1 \oplus P_2)}}. \tag{3.7}$$

Equation (3.7) computes the number of standard deviations away from the expected number of occurrences under the null model. This significance score can be calculated using the aggregate counts of candidate phrases, which can be efficiently obtained from the frequent phrase-mining algorithm. This significance score can be considered a generalization of the t-statistic which has been used to identify dependent bigrams [Church et al., 1991; Pedersen, 1996]. Since the null hypothesis is a simple independence assumption, the binary hypothesis testing to accept or reject the null hypothesis is trivial. Instead, we use the significance score as a concordance measure by which to guide the algorithm in selecting phrases to merge. A high significance indicates a high-belief that two phrases are highly associated and should be merged.

As all merged phrases are frequent phrases, we have fast access to the aggregate counts and they can be used to calculate the significance values for each potential merging. By using proper data structures such as a maximum heap, the contiguous pair with the highest significance can be selected and merged in logarithmic time, $\mathcal{O}(log(l_d))$ for each document. This complexity can once again be reduced by segmenting each document into smaller chunks according to phrase-invariant punctuation.

3.3.3 TOPICAL PHRASE RANKING

After the set of frequent phrases of mixed lengths is mined, they should be ranked with regard to the representativeness of each topic in the hierarchy, based on the popularity and purity criteria mentioned in Section 3.1.

We use the topic word distributions inferred from our model to estimate the topical count $c_{i,P}(t)$ of each phrase P in i-th document :

$$c_{i,P}(t) = c_{i,P}(\pi_t)p(t|P,\pi_t) = c_{i,P}(\pi_t)\frac{\rho_{\pi_t,\chi_t}\prod_{v\in P}\phi_{t,v}}{\sum_{z=1}^{C_{\pi_t}}\rho_{\pi_t,z}\prod_{v\in P}\phi_{\pi_t\odot z,v}}. \tag{3.8}$$

Let the conditional probability $p(P|t)$ be the probability of "randomly choose a document and a phrase that is about topic t, the phrase is P." It can be estimated as $p(P|t) = \frac{1}{D}\sum_{i=1}^{D}\frac{c_{i,P}(t)}{\sum_{P'}c_{i,P'}(t)}$. The popularity of a phrase in a topic t can be quantified by $p(P|t)$. The purity can be measured by the log ratio of the probability $p(P|t)$ conditioned on topic t, and the probability $p(P|\pi_t)$ conditioned on its parent topic π_t: $\log\frac{p(P|t)}{p(P|\pi_t)}$.

As in KERT, a good way to combine these two factors is to use their product:

$$r_t(P) = p(P|t)\log\frac{p(P|t)}{p(P|\pi_t)} \tag{3.9}$$

which has an information-theoretic meaning: the pointwise KL-divergence between two probabilities. Finally, we use $(1-\omega)r_t(P) + \omega p(P|t)\log sig(P)$ to rank phrases in topic t in the descending order.

3.4 EMPIRICAL ANALYSIS

3.4.1 THE IMPACT OF THE FOUR CRITERIA

In the first set of experiments, we use the DBLP dataset - a collection of paper titles in Computer Science - to evaluate the effect of the four criteria to find topical phrases that appear to be high quality to human judges, via a user study.

Ranking Methods for Comparison
To evaluate the performance of KERT, we compare several variations of the phrase ranking function, as well as two baseline ranking functions.

The variations of KERT are constructed with each of the four-phrase quality criteria ignored in turn. We refer to these versions as KERT–pop, KERT–pur, KERT–con, and KERT–com. These variations nicely represent the possible settings for the parameters γ and ω, which are described in Section 3.2. KERT–com sets $\gamma = 0$ to demonstrate what happens when we retain all *closed* phrases. As γ approaches 1, more phrases will be filtered but a very small number of *maximal* phrases (no supersets are frequent) will not be. KERT–con sets $\omega = 0$ and KERT–pur sets $\omega = 1$, to demonstrate the effect of ignoring concordance for the sake of maximizing purity, and ignoring purity to optimize for concordance, respectively.

The baselines are from Zhao et al. [2011], who focus on topical keyphrase extraction in microblogs, claiming that their method can be used for other text collections. Their two best performing methods are kpRelInt and kpRel. Their main ranking function kpRelInt considers the heuristics of phrase interestingness and relevance. Their notion of interestingness is represented by re-Tweets which is unavailable in our dataset. We replace the interestingness measure with the relative frequency of the phrase in the dataset, and refer to this method as kpRelInt*. kpRel considers only the relevance heuristic, which is similar to our notion of popularity. Neither methods model the notion of concordance or completeness.

Qualitative Results

Table 3.3 shows the top 10 ranked topical phrases generated by each method for the topic of Machine Learning. kpRel and kpRelInt* yield very similar results, both clearly favoring unigrams. Removing popularity from our ranking function yields the worst results, confirming the intuition that a high quality phrase must at minimum have good popularity. Without purity, the function favors bigrams and trigrams that all seem to be meaningful, although several high quality unigrams such as "learning" and "classification" no longer appear. Removing concordance, in contrast, yields meaningful unigrams but very few bigrams, and looks quite similar to the kpRelInt* baseline. Finally, without completeness, phrases such as "support vector" and "vector machines" are improperly highly ranked, as both are sub-phrases of the high quality trigram "support vector machines."

User Study and Quantitative Results

To quantitatively measure phrase quality, we invite people to judge the generated topical phrases generated by the different methods. There are four topics which are clearly interpretable as Machine Learning, Databases, Data Mining, and Information Retrieval. For each of the four topics, we retrieve the top 20 ranked phrases by each method. These phrases are gathered together per topic and presented in random order. Users are asked to evaluate the quality of each phrase on a 5 point Likert scale.

To measure the performance of each method given the user study results, we adapt the **nKQM@K measure** (normalized phrase quality measure for top-K phrases) from Zhao et al. [2011], which is itself a version of the *nDCG* metric from information retrieval [Järvelin and

Table 3.3: Top 10 ranked keyphrases in the Machine Learning topic by different methods

Method	Top 10 Topical Keyphrases
kpRelInt*	learning / classification / selection / models / algorithm / feature / decision / bayesian / trees / problem
kpRel	learning / classification / learning classification / selection / selection learning / feature / decision / bayesian / feature learning / trees
KERT$_{-cov}$	effective / text / probabilistic / identification / mapping / task / planning / set / subspace / online
KERT$_{-pur}$	support vector machines / feature selection / reinforcement learning / conditional random fields / constraint satisfaction / decision trees / dimensionality reduction / constraint satisfaction problems / matrix factorization / hidden markov models
KERT$_{-phr}$	learning / classification / selection / feature / decision / bayesian / trees / problem / reinforcement learning / constraint
KERT$_{-com}$	learning / support vector machines / support vector / reinforcement learning / feature selection / conditional random fields / vector machines / classification / support machines / decision trees
KERT	learning / support vector machines / reinforcement learning / feature selection / conditional random fields / classification / decision trees / constraint satisfaction / dimensionality reduction / matrix factorization

Kekäläinen, 2002]. We define nKQM@K for a method M using the top-K generated phrases:

$$nKQM@K = \frac{1}{k} \sum_{t=1}^{k} \frac{\sum_{j=1}^{K} \frac{score_{aw}(M_{t,j})}{\log_2(j+1)}}{IdealScore_K},$$

where T is the number of topics, and $M_{t,j}$ refers to the jth phrase generated by method M for topic t. $IdealScore_K$ is calculated using the scores of the top K phrases out of all judged phrases.

Table 3.4 compares the performance across different methods. The top performances are clearly variations of KERT with different parameter settings. As expected, KERT–pop exhibits the worst performance. The baselines perform slightly better, and it is interesting to note that kpRel, which is smoothed purity, performs better than kpRelInt*, and even slightly better than KERT–con. This is because kpRelInt* adds in a measure of *overall* phrase popularity in the entire collection, which hurts rather than helps for this task. KERT–pur performs the best—which may reflect human bias towards longer phrases—with an improvement of at least 50% over the kpRelInt* baseline for all reported values of K.

Table 3.4: nKQM@K (methods ordered by performance)

Method	nKWM@5	nKWM@10	nKWM@20
KERT$_{-pop}$	0.2605	0.2701	0.2448
kpRelInt*	0.3521	0.3730	0.3288
KERT$_{-con}$	0.3632	0.3616	0.3278
kpRel	0.3892	0.4030	0.3767
KERT$_{-com}$	**0.5124**	**0.4932**	**0.4338**
KERT	**0.5198**	**0.4962**	**0.4393**
KERT$_{-pur}$	**0.5832**	**0.5642**	**0.5144**

Maximizing Mutual Information

In order to perform an objective evaluation, we use a dataset which, unlike the DBLP collection, has been labeled. In the arXiv dataset, each physics paper title is labeled by its authors as belonging to the subfield of Optics, Fluid Dynamics, Atomic Physics, Instrumentation and Detectors, or Plasma Physics. Since the titles are labeled, we can explore which method maximizes the mutual information between phrase-represented topics and titles. As the collection has 5 categories, we set k=5.

For each method, we do multiple runs for various values of K (the number of top-ranked phrases per topic considered), and calculate the mutual information MI_K for that method as a function of K. To calculate MI_K, we label each of the top K phrases in every topic with the topic in which it is ranked highest. We then check each paper title to see if it contains any of these top phrases for each topic. Finally, we compute mutual information at K:

$$MI_K = \sum_{t,c} p(t,c) \, \log_2 \frac{p(t,c)}{p(t)p(c)},$$

where c is the Primary Category label, and t is the topic indicated by topical phrases.

We compare the baselines (kpRelInt* and kpRel), KERT, and variations of KERT where only popularity (KERTpop), only purity (KERTpur), and only coverage and purity (KERTpop+pur) are used in the ranking function. Figure 3.2 shows MI_K for each method for a range of K.

It is clear that for MIK, popularity is more important than purity, since KERTpur is by far the worst performer. Both baselines perform nearly as well as KERTpop, and all are comfortably beaten by KERTpop+pur (> 20% improvement for K between 100 and 600), which uses both popularity and purity measure. It is interesting to note that adding in the concordance and completeness measures yields no improvement in MIK. However, the experiments with the DBLP dataset demonstrate that these measures—particularly concordance—are very helpful in the eyes of expert human judges. In contrast, while MIK is definitely improved with the addition of the purity measure, people seem to prefer that this metric not affect the phrase ranking. These

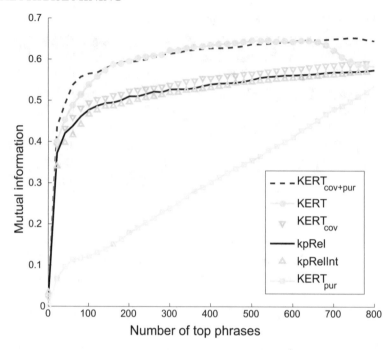

Figure 3.2: Mutual information at K (MIK) for various K. Methods in legend are ordered by performance, high to low.

observations show interesting differences between theory-based and human-centric evaluation metrics.

3.4.2 COMPARISON OF MINING METHODS

We compare the following five methods.

- TNG [Wang et al., 2007]—A state-of-the-art approach to n-gram topic modeling. By using additional latent variables to model bi-grams and adding word-specific multinomials, TNG can be used to construct topical phrases.

- TurboTopics [Blei and Lafferty, 2009]—A post-processing algorithm to LDA. It leverages permutation tests and a back-off N-Gram language model to recursively merge same-topic terms from LDA into more understandable groupings.

- PD-LDA [Lindsey et al., 2012]—A hierarchical topical model that infers both phrases and topics. Using hierarchical Pitman-Yor processes, the topic is naturally shared to all constituents in a phrase.

- KERT—Described in Section 3.2.

- ToPMine—Described in Section 3.3.

Datasets

- **DBLP titles**. A set of titles of computer science papers. The collection has 1.9M titles, 152K unique words, and 11M tokens.

- **20Conf**. Titles of papers published in 20 conferences related to the areas of Artificial Intelligence, Databases, Data Mining, Information Retrieval, Machine Learning, and Natural Language Processing. It contains 44K titles, 5.5K unique words, and 351K tokens.

- **DBLP abstracts**. Computer science abstracts containing 529K abstracts, 186K unique words, and 39M tokens.

- **TREC AP news**. News dataset (1989) containing 106K full articles, 170K unique words, and 19M tokens.

- **ACL abstracts**. ACL abstracts containing 2K abstracts, 4K unique words, and 231K tokens.

- **Yelp Reviews**. Yelp reviews containing 230K Yelp reviews and 11.8M tokens.

Interpretability

We present two user studies to evaluate the quality of mined phrases.

First, we use an *intrusion detection* task to evaluate topical separation. The intrusion detection task involves a set of questions asking humans to discover the "intruder" object from several options (see Section 2.3.2). The results of this task evaluate how well the phrases are separated in different topics.

For each method, we sample 20 Phrase Intrusion questions, and asked three annotators to answer each question. Figure 3.3 shows the average number of questions that is answered "correctly" (matching the method).

The second task is motivated by our desire to extract high-quality topical phrases and provide an interpretable visualization. This task evaluates both topical coherence on the full topical phrase list and phrase quality. We first visualize each algorithm's topics with lists of topical phrases sorted by topical frequency. For each dataset, five domain experts are asked to analyze each method's visualized topics and score each topical phrase list based on two qualitative properties:

- **Topical coherence:** We define topical coherence as homogeneity of a topical phrase list's thematic structure. This homogeneity is necessary for interpretability. We ask domain experts to rate the coherence of each topical phrase list on a scale of 1 *to* 10.

- **Phrase quality:** To ensure that the phrases extracted are meaningful and not just an agglomeration of words assigned to the same topic, domain experts are asked to rate the quality of phrases in each topic from 1 *to* 10.

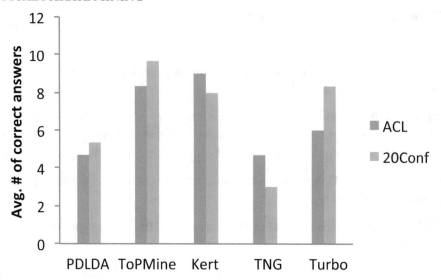

Figure 3.3: Phrase intrusion task.

For each expert, ratings are standardized to a z-score. We compute each algorithm's topical scores by averaging those of five experts. The results are shown in Figures 3.4 and 3.5.

From Figures 3.3 and 3.4 we can tell that TopMine achieves similar performance to KERT in phrase intrusion, and demonstrates the best performance in topical coherence and phrase quality. This is mainly because KERT is designed for short text. A surprising occurrence is TNG and PD-LDA's poor performance in phrase intrusion. We suspect that this may be due to the many hyperparameters these complex models rely on and the difficulty in tuning them. In fact, the authors of PD-LDA make note that two of their parameters have no intuitive interpretation. Finally, Turbo Topics demonstrates above average performance on both datasets and user studies; this is likely a product of the rigorous permutation test the method employs to identify key topical phrases.

3.4.3 SCALABILITY

To understand the run-time complexity of our topical phrase mining methods, we need to couple the phrase mining with certain topic modeling method in order to have a fair comparison with the other methods that integrated topic mining and phrase mining. We use a background LDA [Danilevsky et al., 2014] for KERT and a phrase-constrained LDA [El-Kishky et al., 2015] for ToPMine. For example, the phrase-constrained LDA takes the "bag-of-phrases" output from ToPMine as constraints in LDA. Figure 3.6 demonstrates the disparity in runtime between the phrase mining and topic modeling portions of ToPMine. Displayed on a log-scale for ease of interpretation we see that the runtime of ToPMine scales linearly as we increase the number of

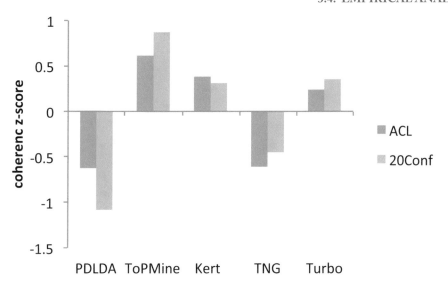

Figure 3.4: Coherence of topics. Results are normalized into z-scores and averaged.

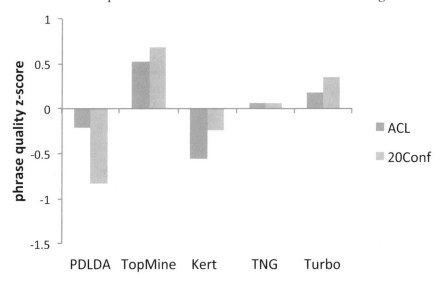

Figure 3.5: Phrase quality. Results are normalized into z-scores and averaged.

documents (abstracts from our DBLP dataset). In addition, one can easily note that the phrase mining portion is of negligible runtime when compared to the topic modeling portion of the algorithm.

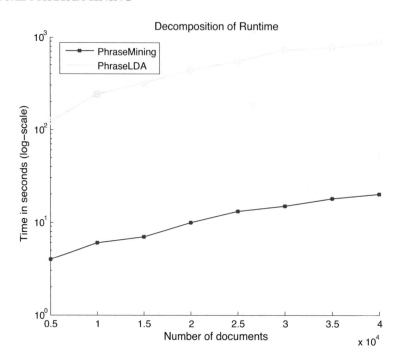

Figure 3.6: The runtime of phrase mining and phrase-constrained topic modeling. The plot above, which is displayed on a log-scale, demonstrates the speed of the phrase-mining portion. With 10 topics and 2000 Gibbs sampling iterations, the runtime of the topic modeling portion is consistently 40X the phrase mining.

To evaluate the scalability, we test the runtime (on the same hardware) for datasets of various sizes and domains. For some datasets, not all methods could not be evaluated due to computational complexity leading to intractable runtimes or due to large memory requirements. We estimate the runtime based on a smaller number of iterations when the computational intractability of an algorithm prevents full evaluation on a specific dataset. The runtime for KERT and ToPMine is the **full runtime** including both phrase mining and topic modeling.

Table 3.5 shows the runtime of each method on our datasets. As expected, complex generative models such as PD-LDA display intractable runtimes outside small datasets showing several magnitudes larger runtime than all methods except Turbo Topics. Turbo Topics displays a similar runtime due to the computationally intensive permutation tests on the back-off n-gram model. These methods were only able to run on the two sampled datasets and could not be applied to the full (larger) datasets. On short documents such as titles, KERT shows great scalability to large datasets barely adding any computational costs to LDA. ToPMine is capable of running on the full DBLP abstracts dataset with runtime in the same order as LDA. Under careful observation, phrase-constrained LDA often runs in shorter time than LDA. This is because phrase-constrained LDA samples a topic once for an entire multi-word phrase, while LDA samples a topic for each word.

Table 3.5: The runtimes of on various datasets of different sizes from different domains. 50 K DBLP titles and 20 K DBLP abstracts are sampled to provide datasets that the state-of-the-art methods can perform on. For instances labeled *, the runtime is estimated by calculating the runtime for one topic and extrapolating for k topics. For instances labeled ∼, runtime is extrapolated from a tractable number of iterations. For instances labeled *, the algorithm could not be applied to the dataset because it exceeded memory constraints (greater than 40 GB) during runtime

Method	sampled DBLP titles (k=5)	DBLP titles (k=30)	sampled DBLP abstracts	DBLP abstracts
PDLDA	3.72(hrs)	∼20.44(days)	1.12(days)	∼95.9(days)
Turbo Topics	6.68(hrs)	>30(days)*	>10(days)*	>50(days)*
TNG	146(s)	5.57(hrs)	853(s)	NA†
LDA	**65(s)**	3.04(hrs)	353(s)	13.84(hrs)
KERT	68(s)	3.08(hrs)	1215(s)	NA†
ToPMine	67(s)	**2.45(hrs)**	**340(s)**	**10.88(hrs)**

Tables 3.6, 3.7, 3.8 are sample results of TopMine on three relatively large datasets—DBLP abstracts, AP News articles, and Yelp reviews. ToPMine was the only method capable on mining these three large, long-text datasets. For comparison, we present the most probable unigrams from phrase-constrained LDA as well as the most probable phrases below the unigrams. Automatic unstemming was performed as a post-processing step to visualize phrases in their most interpretable form. In many cases we see uninterpretable unigram topics that are made easier to interpret with the inclusion of topical phrases. Overall, we can see that for datasets that naturally form topics such as events in the news and computer science subareas, ToPMine yields high quality topical phrases. For noisier datasets such as Yelp, we find coherent, yet lower quality topical phrases.

Table 3.6: Five topics from a 50-topic run of ToPMine on the full DBLP abstracts dataset. Topics can be interpreted as search/optimization, NLP, Machine Learning, Programming Languages, and Data Mining

Topic 1	Topic 2	Topic 3	Topic 4	Topic 5
problem	word	data	programming	data
algorithm	language	method	language	patterns
optimal	text	algorithm	code	mining
solution	speech	learning	type	rules
search	system	clustering	object	set
solve	recognition	classification	implementation	event
constraints	character	based	system	time
programming	translation	features	compiler	association
heuristic	sentences	proposed	java	stream
genetic	grammar	classifier	data	large
genetic algorithm	natural language	data sets	programming language	data mining
optimization problem	speech recognition	support vector machine	source code	data sets
solve this problem	language model	learning algorithm	object oriented	data streams
optimal solution	natural language processing	machine learning	type system	association rules
evolutionary algorithm	machine translation	feature selection	data structure	data collection
local search	recognition system	paper we propose	program execution	time series
search space	context free grammars	clustering algorithm	run time	data analysis
optimization algorithm	sign language	decision tree	code generation	mining algorithms
search algorithm	recognition rate	proposed method	object oriented programming	spatio temporal
objective function	character recognition	training data	java programs	frequent itemsets

Table 3.7: Five topics from a 50-topic run of ToPMine on a large collection of AP News articles(1989). Topics can be interpreted as environment, Christianity, Palestine/Israel conflict, Bush Administration (Senior), and health care

Topic 1	Topic 2	Topic 3	Topic 4	Topic 5
plant	church	palestinian	bush	drug
nuclear	catholic	israeli	house	aid
environmental	religious	israel	senate	health
energy	bishop	arab	year	hospital
year	pope	plo	bill	medical
waste	roman	army	president	patients
department	jewish	reported	congress	research
power	rev	west	tax	test
state	john	bank	budget	study
chemical	christian	state	committee	disease
energy department environmental protection agency	roman catholic	gaza strip	president bush	health care
nuclear weapons	pope john paul	west bank	white house	medical center
acid rain	john paul	palestine liberation organization	bush administration	united states
nuclear power plant	catholic church	united states	house and senate	aids virus
hazardous waste	anti semitism	arab reports	members of congress	drug abuse
savannah river	baptist church	prime minister	defense secretary	food and drug administration
rocky flats	united states	yitzhak shamir	capital gains tax	aids patient
nuclear power	lutheran church	israel radio	pay raise	centers for disease control
natural gas	episcopal church	occupied territories	house members	heart disease
	church members	occupied west bank	committee chairman	drug testing

Table 3.8: Five topics from a 10-topic run of ToPMine on the Yelp reviews dataset. Quality seems to be lower than the other datasets, yet one can still interpret the topics: breakfast/coffee, Asian/Chinese food, hotels, grocery stores, and Mexican food

Topic 1	Topic 2	Topic 3	Topic 4	Topic 5
coffee	food	room	store	good
ice	good	parking	shop	food
cream	place	hotel	prices	place
flavor	ordered	stay	find	burger
egg	chicken	time	place	ordered
chocolate	roll	nice	buy	fries
breakfast	sushi	place	selection	chicken
tea	restaurant	great	items	tacos
cake	dish	area	love	cheese
sweet	rice	pool	great	time
ice cream	spring rolls	parking lot	grocery store	mexican food
iced tea	food was good	front desk	great selection	chips and salsa
french toast	fried rice	spring training	farmer's market	food was good
hash browns	egg rolls	staying at the hotel	great prices	hot dog
frozen yogurt	chinese food	dog park	parking lot	rice and beans
eggs benedict	pad thai	room was clean	wal mart	sweet potato fries
peanut butter	dim sum	pool area	shopping center	pretty good
cup of coffee	thai food	great place	great place	carne asada
iced coffee	pretty good	staff is friendly	prices are reasonable	mac and cheese
scrambled eggs	lunch specials	free wifi	love this place	fish tacos

CHAPTER 4

Entity Topical Role Analysis

People and other entities are often characterized by the topics and themes they are working on, communicating about and involved in. The roles played by different entities in these topics are of great interest in many contexts of analysis. We may be interested in discovering the role of an author in a research community, or the contribution of a user to a social network community organized around similar interests. These types of role discovery tasks center around topical communities mined from text-attached information networks.

We are also often interested in analyzing such roles at different levels of granularity. In the real world, topical communities—communities built around shared topics—are naturally hierarchical. People participate in large communities, encompassing many interests, as well as small, focused subcommunities. Therefore, in order to analyze the various roles that an entity plays in such different contexts, we must also be able to work with topical communities and subcommunities.

In this section we study mining entity roles in hierarchical topical communities. The topical communities are discovered using the methods presented and visualized in previous two chapters. We can then discover the roles of authors who publish in these communities. For example, in the context of computer science topics, the community centered around topics on query processing and optimization may be described by the phrases {"query processing," "query optimization,"...}, while its parent community on general database topics may be described by {"query processing," "database systems," "concurrency control,"...}. The hierarchical structure of the topical communities allows us to distinguish between, e.g., authors who publish on a diverse range of database topics, and authors who are particular experts in query processing.

4.1 ROLE OF GIVEN ENTITIES

This section focuses on the type-A question: Given a set of entities E, what are their roles?

We introduce two ways to analyze a specific entity's role in a given topic.

4.1.1 ENTITY SPECIFIC PHRASE RANKING

In order to specify an entity or several entities' role in a topic, we want to highlight the specific phrases which illustrate the contribution of them. We now therefore introduce an *entity specific phrase* ranking function:

$$r(P|t, E) = -p(P|t)log(\frac{p(P|t)}{p(P|t, E)}), \tag{4.1}$$

where E can be a set of entities in general. It has a nice information theoretic interpretation as the pointwise Kullback-Leibler (KL) divergence between the likelihood of seeing phrase P in the documents in topic t, and the likelihood of seeing phrase P specifically in the documents linked to entity set E, in t. Pointwise, KL divergence is a distance measure between two probabilities. Therefore, $r(P|t, E)$ upranks P if its frequency in the topic in conjunction with the entity set E is higher than would be expected, based on its overall topical frequency. $p(P|t)$ can be obtained from the topic discovery and phrase mining described in previous two chapters. We can use Bayes rule to estimate $p(P|t, E)$:

$$p(P|t, E) = \frac{p(P, E|t)}{p(E|t)} = \frac{f_t(P \cup E) \prod_{v \in P} \phi_{t,v}}{f_t(E) \sum_z \in [C_{\pi_t}] \prod_{v \in P} \phi_{\pi_t \odot z, v}}.$$

Using only the entity specific phrase ranking does not give ideal results. Table 4.1 shows the roles of two authors, Philip S. Yu and Christos Faloutsos, in one of the subcommunities of Data Mining subtopics. Using only the contribution ranking function defined in Eq. (4.1) results in poor quality phrases such as 'fast large.' On the other hand, using only the phrase quality ranking function defined in Chapter 3—which we refer to here as $r(P|t)$—is also insufficient, as it only evaluates the quality of a phrase, regardless of any entity information. Therefore, we define a *combined* ranking function for a phrase P which incorporates both the relationship between the entity E and the phrase, as well as phrase quality:

$$Comb(P|t, E) = \alpha r(P|t, E) + (1 - \alpha)r(P|t). \tag{4.2}$$

Table 4.1 illustrates that the combined ranking function yields a better list of phrases to represent the roles of the authors, with $\alpha = 0.5$.

4.1.2 DISTRIBUTION OVER SUBTOPICS

Another way to analyze the entity roles is to examine the subtopical frequency $f_{t\odot z}(E)$, the number of documents attributed to E in topic t. Denote by $\mathcal{D}_E$ the set of all documents connected to E. For example, in the DBLP dataset, the subset $\mathcal{D}_A$ is the set of papers authored by A, and $\mathcal{D}_V$ is the set of papers published in V.

$f_t(E)$ can be caluclated by summing up the contributions of all the documents $d_E \in \mathcal{D}_E$ to topic t:

$$f_t(E) = \Sigma_{d_E \in \mathcal{D}_E} f_t(d_E). \tag{4.3}$$

Table 4.1: Using phrase quality, entity specific phrase ranking, and a combination of both to represent the roles of Philip S. Yu and Christos Faloutsos in a Data Mining subcommunity

Phrase Quality	P.S. Yu (entity specific)	C. Faloutsos (entity specific)	P.S. Yu (combined)	C. Faloutsos (combined)
time series	data indexing	time warping	time series	nearest neighbor
nearest neighbor	data similarity	distance	nearest neighbor	time warping
moving objects	distance	fast time	time series data	moving objects
time series data	fast large	time similarity	moving objects	nearest neighbor search
nearest neighbor queries	similarity indexing	fast large	time series mining	time series
mining time series	time series patterns	fast similarity	time series patterns	distance

For a single entity e, if the topic model generates a multinomial distribution ϕ^x for each entity type x, we can easily estimate the subtopical frequency using Bayes rule:

$$f_{t\odot z}(e) = f_t(e) \frac{\phi^x_{t\odot z,e}\rho_{t\odot z}}{\sum_{c\in[C_t]} \phi^x_{t\odot c,e}\rho_{t\odot c}}. \tag{4.4}$$

Otherwise, we provide the following heuristic method to estimate entity topical frequency. We must first estimate the topical frequency of every document $d_E \in \mathcal{D}_E$. In Section 3.2.2 we described how to estimate $f_t(P)$, the topical frequency of phrase P. We proceed in a similar top-down recursive fashion in order to estimate the *document* topic frequency, $f_t(d_E)$.

For each document d_E we first perform the intermediate step of calculating the *total phrase frequency* of d_E in each topic by adding up the normalized topical frequencies of all the phrases in d_E:

$$TPF_{t\odot z}(d_E) = \sum_{P\in d_E} \frac{f_{t\odot z}(P)}{\sum_{c\in[C_t]} f_{t\odot c}(P)}. \tag{4.5}$$

The next step is to calculate the document topical frequency of d_E recursively:

$$f_{t\odot z}(d_E) = \frac{TPF_{t\odot z}(d_E)}{\sum_{c\in[C_t]} TPF_{t\odot c}(d_E)} f_t(d_E). \tag{4.6}$$

The topic frequency of a document is distributed among that topic's children, so that the document frequency in a given topic is the sum of the document frequencies in the topic's children, $\sum_{c \in [C_t]} f_{t \odot c}(d) = f_t(d)$. One exception is that a few documents may contain no frequent topical phrases in any subcommunities because we filter out infrequent topical phrases. For such documents we do not count their contribution to any subcommunities.

Figure 4.1 shows a hypothetical distribution of document frequency for some document. The document frequency values for every set of subcommunities sum up to the document frequency in the parent topic. The frequency at the root is 1 for every document.

Since some documents may not contribute to any of the subtopics, the entity frequency in a given topic should be equal to or larger than the sum of the entity frequencies in the topic's children, $\sum_{c \in C_t} f_{t \odot c}(E) \le f_t(E)$.

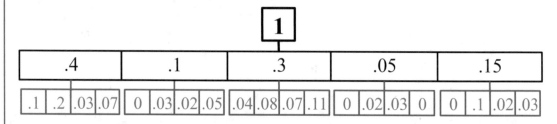

Figure 4.1: A hypothetical distribution of document frequency values for a document, in a hierarchy with three levels, beginning at the root.

4.1.3   CASE STUDY

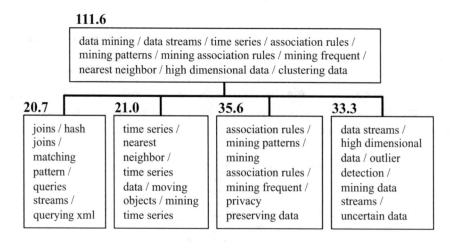

Figure 4.2: The roles of Philip S. Yu in Data Mining.

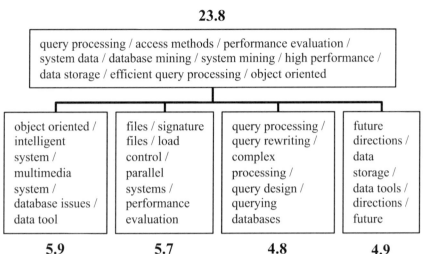

23.8

query processing / access methods / performance evaluation / system data / database mining / system mining / high performance / data storage / efficient query processing / object oriented

| object oriented / intelligent system / multimedia system / database issues / data tool | files / signature files / load control / parallel systems / performance evaluation | query processing / query rewriting / complex processing / query design / querying databases | future directions / data storage / data tools / directions / future |

5.9   **5.7**   **4.8**   **4.9**

Figure 4.3: The roles of Christos Faloutsos in Data Mining.

As an example, Figures 4.2 and 4.3 show the roles of Christos Faloutsos and Philip S. Yu in the Data Mining topic, and its subtopic. We also show the entity frequency for each topic ($f_t(E)$), which represents the estimate for the number of papers written by that author in the topic. The sum of the entity frequencies in the subtopic do not quite add up to the entity frequency of the parent topic because, as discussed in Section 4.1.2, a document does not contribute to the child subtopics if all of its phrases have become too infrequent in them.

While both authors are prominent in the Data Mining topic, Figures 4.2 and 4.3 illustrate how their roles are contrasted in that topic, and even more strongly in the subtopic. For instance, in the third (from left) subtopic, Philip Yu contributes work on the topics of mining frequent patterns and association rules, whereas the contribution of Christos Faloutsos is more geared towards the topics of mining large datasets and large graphs.

As another example of role discovery, Figure 4.4 shows the role of the SIGIR venue in all five top level topics, as well as the subtopic of Machine Learning and Information Retrieval. The role of a venue in a topic is represented by those topics within the topic that are published in the venue. Thus, we can see that the Machine Learning topics that get published in SIGIR are techniques related to IR tasks such as feature selection methods that may be used for filtering, and approaches to text categorization and classification problems.

By examining the roles of different venues within a single topic, we can also gain some insight to the flavor of each venue. As an example, Table 4.2 compares the roles of three venues— SIGIR, WWW, and ECML—in the general IR topic. While both SIGIR and WWW are usually characterized as IR venues, we can clearly see that SIGIR plays a more broad role, publishing most of the topics present in the topic, whereas WWW focuses only on those topics that are

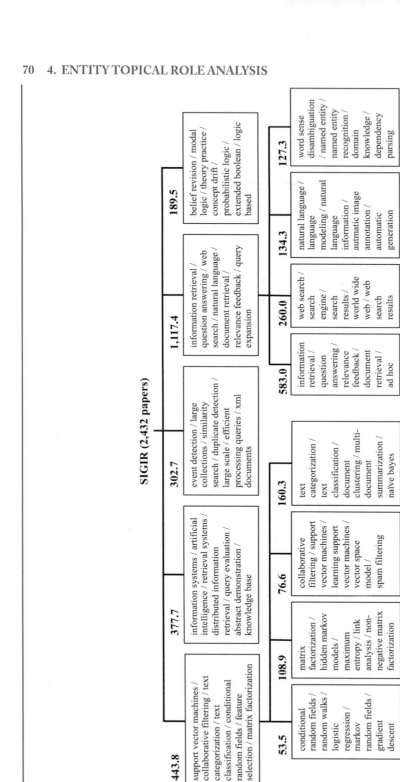

Figure 4.4: The role of the venue SIGIR in several topics and subtopics. The estimated number of papers published in SIGIR within each topic is also shown.

directly related to the web. On the other hand, ECML is considered to be an ML venue, and its contribution to the IR topic is the publishing of papers on topics that use machine learning techniques. Note that all three venues share some high-ranked phrases, illustrating how the roles of all three venues overlap in this topic. If we were to strictly label venues, and therefore the papers they publish, as belonging exclusively to one or another topic, we would not be able to discover these interesting roles.

Table 4.2: The roles of three venues—SIGIR, WWW, and ECML—in the general Information Retrieval topic

SIGIR	WWW	ECML
information retrieval	web search	word sense disambiguation
question answering	semantic web	world wide web
web search	search engine	information extraction
natural language	question answering	semantic role labeling
document retrieval	web pages	knowledge discovery
relevance feedback	world wide web	query expansion
query expansion	web services	machine translation

4.2   ENTITIES OF GIVEN ROLES

The role of an entity in a topical community can be interpreted as that entity's contribution to the community. For example, the role of an author is represented by the work the author does on the topics; the role of a venue is represented by the topics which get published in the venue. Therefore, a natural question to ask is which entities play the roles of top contributors to a particular topical community.

In principle, we can rank entities according to entity popularity $p(e|t)$ for a topic t. However, this entity ranking function is not able to discover entities who are more dedicated to their role in a given topic than to sibling communities. In order to take this into account, we adapt the notion of purity, as introduced in Section 3.1, to apply to entities.

We evaluate the purity of entity e in t by comparing the probability of seeing a entity e conditioned on topic t and the contrastive probability of seeing it in a mixture of topics t and t'.

The criteria of entity purity and popularity can then be unified in an analogous way to Section 3.2. We refer to ranking entities by this value as $ERank_{Pop+Pur}$:

$$ERank_{Pop+Pur}(e,t) = p(e|t) \log \frac{p(e|t)}{p(e|t,t')}.$$

Table 4.3 shows the top ranked authors in the four subtopics of Data Mining, based on $p(e|t)$ and $ERank_{Pop+Pur}$. When only popularity is used for ranking, many authors are highly ranked in all topics (e.g., Philip Yu, Jiawei Han, and Christos Faloutsos are top-5 authors in every topic). When both popularity and purity criteria are taken into account, only those authors who are significantly more dedicated to one topic are highly ranked, resulting in no overlap between topics. Some prolific authors, such as Christos Faloutsos, are no longer highly ranked anywhere, because their contributions are fairly equal among the topics. We are able to easily discover both of these roles.

Table 4.3: Top-ranked authors in the four subtopics of Data Mining, based on popularity only, and popularity + purity

{sensor networks, selectivity estimation, large databases, pattern matching, spatiotemporal moving objects, large collections}	*{time series, nearest neighbor, moving objects, time series data, nearest neighbor queries}*	*{association rules, large scale, mining association rules, privacy preserving, frequent itemsets}*	*{high dimensional, data streams, data mining, high dimensional data, outlier detection}*
divesh srivasta nick koudas jiawei han philip s. yu christos faloutsos	eamonn j. keogh philip s. yu christos faloutsos hans-peter kriegel jiawei han	jiawei han philip s. yu jian pei christos faloutsos ke wang	philip s. yu jiawei han charu c. aggarwal jian pei christos faloutsos
divesh srivasta surat chaudhiri nick koudas jeffrey f. naughton yannis papakonstantinou	eamonn j. keogh jessica lin michail vlachos michael j. passani matthias renz	jiawei han ke wang xifeng yan bing liu mohammed j. zaki	charu c. aggarwal graham cormode s. muthukrishnan philip s. yu xiaolei li

Table 4.4 shows further examples of ranking authors (using $ERank_{Cov+Pur}$) within two sub-communities of the Database community. By showing the top-ranked phrases for each author in a community (we discussed how these are generated in the last section) we are able to see both which authors play the most important roles, and what part of the community each author contributes to.

Table 4.4: The top-ranked authors (using $ERank_{Pop+Pur}$) in two subtopics of the Database topic, along with each author's top-ranked phrases in each topic. Each subtopic is represented by its top-ranked phrases, shown in the first row of each table

{query processing / query optimization / deductive databases / materialized views / microsoft sql server / relational databases}	
elke a. rundensteiner	query processing / query optimization / materialized views / stream processing / object-oriented databases
hamid pirahesh	query processing / query optimization / materialized views / relational data / relational xml
surajit chaudhuri	query optimization / relational databases / microsoft sql server / materialized views / relational data
jeffrey f. naughton	materialized views / xml query / query processing / relational xml / maintenance view
per-åke lar-son	materialized views / microsoft sql server / query optimization / materialized maintenance views / relational data
vivek r. narasayya	microsoft sql server / materialized views / relational databases / query management / sql data
serge abiteboul	materialized views / xml data / schemas / query evaluation / materialized maintenance views
{concurrency control / database systems / main memory load shedding / database concurrency control / load balancing}	
avi silberschatz	concurrency control / main memory / locking / database systems / transaction management
david b. lomet	recovery / systems recovery / b-trees / transactions recovery / performance access
henry k. korth	concurrency control / database systems / main memory / protocol / transaction systems
bharat k. bhargava	concurrency control / distributed systems / distributed database / recovery / distributed database systems
c. mohan	concurrency control / recovery / locking / data systems / transaction systems
ahmed k. elmagarmid	database systems / concurrency control / distributed database / distributed systems / access control
nancy lynch	concurrency control / locking / nested transactions / control transactions / concurrency transactions

CHAPTER 5

Mining Entity Relations

In this chapter, we discuss mining latent relations among entities. It is well recognized that different types of relationships have essentially different influence between entities, which forms the subtle force that governs the dynamics of complex networks. For example, in the social network, a graduate's research topic may be mainly influenced by his advisor; while his living habits may be influenced by his family. Awareness of the relationship types can offer abundantly additional information for many applications such as expert finding. For example, if we know advisor-advisee relationships between researchers, we can discover how researchers form different communities, how research topics have been emerging and evolving in the past years, and how a researcher influences the academic research community.

The main challenge for many entity relation mining problems is to model the dependency of multiple relations. We illustrate several modeling techniques in various examples of different domains, including mining academic family, online forum conversation, and social network friend circles.

5.1 UNSUPERVISED HIERARCHICAL RELATION MINING

In the unsupervised setting, we have only a few number of features and constraints to characterize the relations. They are either equally important for relation discovery, or have unequal but known importance.

We take a case study of advisor-advisee relationship. In this section, we use a computer science bibliographic network as an example, to analyze the roles of authors and to discover the likely advisor-advisee relationships. To clearly illustrate the problem, Figure 5.1 gives an example of advisor-advisee relationship mining. The left figure shows the input: an temporal collaboration network, which consists of authors, papers, and paper-author relationships. The middle figure shows the output of our analysis: an author network with solid arrow indicating the advising relationship, and dotted arrow suggesting potential but less probable relationship. For example, the arrow from Bob to Ada indicates that Ada is identified as the advisor of Bob. The triple on the edge, i.e., (0.8, [1999, 2000]), represents Ada has the probability of 80% to be the advisor of Bob from 1999–2000. The right figure gives an example of visualized chronological hierarchies. The parent-child relation in the tree corresponds to the advisor-advisee relationship.

The following subsections describe a probabilistic graphical model solution.

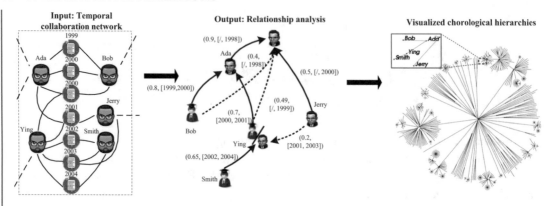

Figure 5.1: Example of advising relationship analysis on publication network.

5.1.1   NOTATIONS

In this subsection, we define notations used throughout this section.

In general, our study takes as input a time-dependent collaboration network $H = \{(\mathcal{V} = \mathcal{V}^p \cup \mathcal{V}^a, \mathcal{E})\}$, where $\mathcal{V}^p = \{p_1, \ldots, p_{n_p}\}$ is the set of publications, with p_i published in time t_i, $\mathcal{V}^a = \{a_1, \ldots, a_{n_a}\}$ is the set of authors, and $\mathcal{E}$ is the set of edges. Each edge $e_{ij} \in \mathcal{E}$ associates the paper p_i and the author a_j, meaning a_j is one author of p_i.

The original heterogeneous network can be transformed into a homogeneous network containing only authors. Let $G = (\mathcal{V}^a \cup \{a_0\}, \mathcal{E}', \{\mathbf{py}_{ij}\}_{e_{ij} \in \mathcal{E}'}, \{\mathbf{pn}_{ij}\}_{e_{ij} \in \mathcal{E}'})$, where a_0 is a virtual author, which will be the root of an advising tree. Each edge $e'_{ij} = (i, j) \in \mathcal{E}$ connects authors a_i and a_j if and only if they have publication together, and there are two vectors associated with the edge, Pub_Year_vector $\mathbf{py}_{ij}$ and Pub_Num_vector $\mathbf{pn}_{ij}$. They are of equivalent length, indicating the year they have publications and the number of coauthored papers they have at that time. For example, $\mathbf{py}_{ij} = (1999, 2000, 2001), \mathbf{pn}_{ij} = (2, 3, 4)$ indicates that author a_i and a_j have coauthored 2, 3, and 4 papers in 1999, 2000, and 2001, respectively. Similarly, we associate with each author two vectors $\mathbf{py_i}$ and $\mathbf{pn_i}$ to, respectively, represent the number of papers and the corresponding published year by author a_i. The two vectors $\mathbf{py_i}$ and $\mathbf{pn_i}$ can be derived from $\mathbf{py_{ij}}$ and $\mathbf{pn_{ij}}$.

We denote the author a_i's advisor as a_{y_i}, where y_i is an introduced hidden variable. If a_i's advisor is a_j, we use $[st_{ij}, ed_{ij}]$ to represent the time interval the advising relation lasts. For brevity we denote $st_i = st_{iy_i}$ and $ed_i = ed_{iy_i}$. If a_i is not advised by anybody in the database, we let $y_i = 0$ to direct a_i's advisor to a virtual node a_0.

In this setting, to find the advisor-advisee relationship, we need not only to decide the value of the hidden variable y_i for each author a_i, but also to estimate the start and the end years st_{iy_i}, ed_{iy_i}. In reality, this problem is more complicated: (i) one could have multiple advisors like master advisors, PhD co-advisors, post-doctorial advisors; (ii) some mentors from industry behave similarly as academic advisors if only judged by the collaboration history; and (iii) one's advisor could be missing in the data set. Therefore, instead of using a boolean model, we adopt a probabilistic model to rank the likelihood of potential advisor(s) for each author. Formally, we denote r_{ij} as the probability of a_j being the advisor of a_i. To reduce the number of authors being ranked, it is beneficial to keep only those potential pairs of advisor-advisee. We construct a subgraph $G' \subset G$ by removing some edges from G and make the remaining edges directed from advisee to potential advisor. Thus, $G' = (\mathcal{V}^a \cup \{a_0\}, \mathcal{E}'_s)$ and $\mathcal{E}'_s \subset \mathcal{E}'$. Later we will show that it is possible to extract a directed acyclic graph (DAG) G' from G. In G', the index set of potential advisors of a given author a_i is denoted $Y_i = \{j | e_{ij} \in \mathcal{E}'_s\}$, e.g., $Y_3 = \{0, 1\}$. Correspondingly, the index set of potential advisees is denoted $Y_i^{-1} = \{j | e_{ji} \in \mathcal{E}'_s\}$.

Then the task becomes finding r_{ij}, st_{ij}, ed_{ij} for every possible advising pair $(i, j) \in \mathcal{E}'_s$. So the output is the DAG $H' = (\mathcal{V}^a \cup \{a_0\}, \mathcal{E}'_s, \{(r_{ij}, st_{ij}, ed_{ij})\}_{(i,j) \in \mathcal{E}'_s})$. After the chronological DAG H' is calculated, the ranking score can be used to predict whether there is an advisor-advisee relationship between every pair of coauthors (a_i, a_j). A simple way to predict is to fetch top k potential advisors of a_i and check whether a_j is one of them while $r_{ij} > r_{i0}$ or $r_{ij} > \theta$, where θ is a threshold such as 0.5. We use $P@(k, \theta)$ to denote this method. It is predictable that large k and large θ leads to better recall and worse precision. If there are training data in the input, they can be used to determine the parameters. If no training data is provided, we can simply use some empirical values, such as the third quartile of all the ranking scores.

5.1.2   ASSUMPTIONS AND FRAMEWORK

Commonsense knowledge is needed for recognizing interesting semantic relationships. Here we make a few general assumptions based on the commonsense knowledge about advisor-advisee relationships.

Assumption 5.1 $\forall 1 \leq x \leq n_a, ed_{y_x} < st_x < ed_x$

This formula reflects the following fact for general consideration of advising relationship. At each time t during the publication history of a node x, x is either being advised or not being advised. Once x starts to advise another node, it will never be advised again. x cannot advise y at the year t_1 if x is advised by any node p at the year t_1. If x advises y, the time y is advised by x is a continuous interval from t_1 to t_2, $t_1 < t_2$. As a result of Assumption 5.1, we need to infer the advisors of all the nodes in the network together, rather than consider them separately. In Section 5.1.4, we will use this assumption in our model.

Assumption 5.2 $\forall 1 \leq x \leq n_a, py^1_{y_x} < py^1_x$

That means for a given pair of advisor and advisee, the advisor always has a longer publication history than the advisee. py^1_x represents the first component of vector $\mathbf{py}_x$. Assumption 5.2 determines that all the authors in the network have a strict order defined by the possible advising relationship. Due to the order, the candidate graph G' is assured to be a DAG. We will use this assumption in the filtering process in Section 5.1.3.

Additional assumptions about the correlation between the potential relationship and the publication history will be discussed in Section 5.1.3. Now we propose a two-stage framework solution for the advisor-advisee relationship mining problem. In stage 1, we preprocess the heterogeneous collaboration network to generate the candidate graph G'. This includes the transformation from H to a homogeneous network G, the construction from G to G', and the estimate of the local likelihood on each edge of G'. In stage 2, these potential relations are further modeled with a probabilistic model. Local likelihood and time constraints are combined in the global joint probability of all the hidden variables. The joint probability is maximized and the ranking score of all the potential relations is computed together. The construction of H' is finished in this stage.

5.1.3 STAGE 1: PREPROCESSING

The purpose of preprocessing is to generate the candidate graph G' and reduce the search space while keeping the real advisor not excluded from the candidate pool in most cases. First, we need to generate according to the collaboration information a homogeneous author network G by processing the papers in the network one by one. For each paper $p_i \in \mathcal{V}^p$, we construct an edge between every pair of its authors and update the vectors **py** and **pn**. The complexity of this process is $\mathcal{O}(\sum_{p_i \in \mathcal{V}^p} \delta_i^2)$, where δ_i is the degree of p_i in H.

Then a filtering process is performed to remove unlikely relations of advisor-advisee. For each edge e_{ij} on G, a_i and a_j has collaboration. To decide whether a_j is a_i's potential advisor, the following conditions are checked. First, Assumption 5.2 is checked. Only if a_j started to publish earlier than a_i, the possibility is considered. Second, some heuristic rules are applied, which are based on the prior intuitive knowledge about advisor-advisee relations.

Here we introduce two measures for the coauthored publications between any pair of collaborators, Kulczinski measure denoted by $kulc$, and imbalance ratio denoted by IR [Wu et al., 2010a]. They are defined as

$$kulc_{ij}^t = \frac{\sum_{py_{ij}^k \le t} pn_{ij}^k}{2} \left(\frac{1}{\sum_{py_i^k \le t} pn_i^k} + \frac{1}{\sum_{py_j^k \le t} pn_j^k} \right) \tag{5.1}$$

$$IR_{ij}^t = \frac{\sum_{py_j^k \le t} pn_j^k - \sum_{py_i^k \le t} pn_i^k}{\sum_{py_i^k \le t} pn_i^k + \sum_{py_j^k \le t} pn_j^k - \sum_{py_{ij}^k \le t} pn_{ij}^k}. \tag{5.2}$$

The Kulczynski measure reflects the correlation of the two authors' publications. Wu et al. [2010a] shows that there usually exists high correlation between the total publications of advisors and advisee. Here we further incorporate the time factor, to calculate the measure year by year, and check whether there is an increase in the sequence $\{kulc_{ij}^t\}_t$. For IR, we calculate the sequences in the same way. IR is used to measure the imbalance of the occurrence of a_j given a_i and the occurrence of a_i given a_j. The intuition is that the advisor has more publications than the advisee during the advising time.

Author a_j is not considered to be a_i's advisor if one of the following conditions holds:

- $R1$: $IR_{ij}^t < 0$ in the sequence $\{IR_{ij}^t\}_t$ during the collaboration period of a_i and a_j;

- $R2$: there is no increase in the sequence $\{kulc_{ij}^t\}_t$ during the collaboration period;

- $R3$: the collaboration period of a_i and a_j lasts only for one year; and

- $R4$: $py_j^1 + 2 > py_{ij}^1$.

When the pair of authors passes the test of selected rules from them, we construct a directed edge from a_i to a_j in G'. In addition, we estimate the starting time and ending time of the advising, as well as the local likelihood of a_j being a_i's advisor l_{ij}. For the estimation we also have various methods. The starting time st_{ij} is estimated as the time they started to collaborate, while the ending time ed_{ij} can be estimated as either the time point when the Kulczynski measure starts to decrease, or the year making the largest difference between the Kulczynski measure before and after it. We refer to the two methods as YEAR1 and YEAR2. And we refer to YEAR as taking the earlier time of the two years estimated by them. After estimating st_{ij} and ed_{ij}, we calculate the average of Kulczynski and IR measure during that period, and use (1) Kulczynski ; (2) IR; and (3) the average of the two as three different definitions of the local likelihood. The last definition is formally

$$l_{ij} = \frac{\sum_{st_{ij} \leq t \leq ed_{ij}} (kulc_{ij}^t + IR_{ij}^t)}{2(ed_{ij} - st_{ij} + 1)}. \tag{5.3}$$

And the other two are similar. The complexity of processing each edge is $\mathcal{O}(\Delta)$, if we assume the oldest paper and the newest one differs Δ in their publication time. The total complexity to transform G to G' is $\mathcal{O}(M\Delta)$, where M is the number of edges in G.

5.1.4   STAGE 2: TPFG MODEL

From the candidate graph G' we know the potential advisors of each author and the likelihood based on local information. By modeling the network as a whole, we can incorporate both structure information and temporal constraint and better analyze the relationship among individual links. Now we define the TPFG model.

For each node a_i, there are three variables to decide: y_i, st_i, and ed_i. Suppose we have already had a local feature function $g(y_i, st_i, ed_i)$ defined on the three variables of any given node. To model the joint probability of all the variables in the network, we define it as the product of all local feature functions.

$$p(\{y_i, st_i, ed_i\}_{a_i \in V^a}) = \frac{1}{Z} \prod_{a_i \in V^a} g(y_i, st_i, ed_i) \tag{5.4}$$

with

$$\forall a_i \in V^a, ed_{y_i} < st_i < ed_i \tag{5.5}$$

where $\frac{1}{Z}$ is the normalizing factor of the joint probability.

Equation (5.5) is the constraint according to Assumption 5.1. To find the most probable values of all the hidden variables, we need to maximize the joint probability of all of them. To estimate the approximate size of the entire search space, assume each author has C candidates and the advising time can vary in a range of Δ, then the combination of all the variables has exponential size $(C\Delta^2)^{na}$. It is intractable to do exhaustive search. We make the first simplification as follows. Suppose a_i and his advisor y_i are given. Instead of letting st_i and ed_i vary, we fix them by optimizing local function $g(y_i, st_i, ed_i)$, i.e.,

$$\{st_i, ed_i\} = arg \max_{st_i < ed_i} g(y_i, st_i, ed_i). \tag{5.6}$$

In this way, st_i and ed_i are tied to the value of y_i. Once y_i is decided, they are derived correspondingly. We can pre-compute the best advising time as st_{ij} and ed_{ij} for each $y_i = j$. Now only $\{y_i\}$ are variables to optimize). If we embed the constraint Eq. (5.5) into the feature function, the objective function becomes

$$p(y_1, \ldots, y_{na}) = \frac{1}{Z} \prod_{i=1}^{na} f_i(y_i | \{y_x | x \in Y_i^{-1}\}) \tag{5.7}$$

with

$$f_i(y_i = j | \{y_x | x \in Y_i^{-1}\}) = g(y_i, st_{ij}, ed_{ij}) \prod_{x \in Y_i^{-1}} I(y_x \neq i \vee ed_{ij} < st_{xi}) \tag{5.8}$$

where

$$I(y_x \neq i \vee ed_{ij} < st_{xi}) = \begin{cases} 1 & y_x \neq i \vee ed_{ij} < st_{xi} \\ 0 & y_x = i \wedge ed_{ij} >= st_{xi} \end{cases} \tag{5.9}$$

is the identity function. If any author a_x is advised by a_i and their advising time conflict, the function takes 0; otherwise it takes 1. In this way the time constraints Eq. (5.5) for all hidden variables are decomposed to many local identity function. Now we only need to optimize Eq. (5.7). We can define $g(y_i, st_{ij}, ed_{ij}) = l_{ij}$ when $y_i = j$. To obtain the rank score of each advising relationship, e.g., a_j advise a_i (shortly $a_j \rightarrow a_i$), we can compute the conditional maximal probability

$$r_{ij} = \max p(y_1, \ldots, y_{na} | y_i = j). \tag{5.10}$$

This simplification assures that for each configuration of $\{y_i\}$, the solution achieves either 0 or the conditional optimum given that configuration. The search space size now becomes C^{na}, reduced but still exponential. Since we have decomposed the dependency of the variables, we can use a factor graph model to accomplish efficient computation.

Figure 5.2 shows a simple TPFG corresponding to the example we have been using so far. The graph is composed of two kinds of nodes: variable nodes and function nodes. Variable nodes map to the hidden variables $\{y_i\}_{i=0}^{na}$. Each variable node corresponds to a function node $f_i(y_i | \{y_x | x \in Y_i^{-1}\})$. All of the edges are of one kind, connecting a variable node with a function

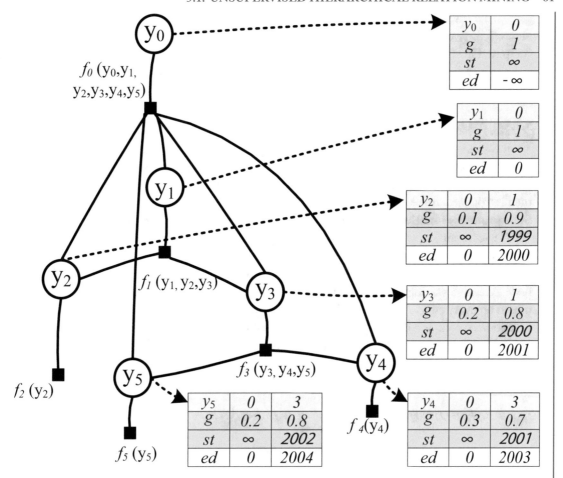

Figure 5.2: Graphical representation of a time-constrained probabilistic factor graph, where $\{y_0, \ldots, y_5\}$ are hidden variables defined on all nodes; $f_i(.)$ represents a factor function defined on a hidden variable and its potential advisee sets as neighbors.

node. There is an edge between one variable node y_x and a function node $f_i(.)$ if and only if $f_i(.)$ depends on y_x. In our case, it is equivalent with $x = i$ or $x \in Y_i^{-1}$ (a.k.a. $i \in Y_x$). The factor graph reflects the dependency of the variables. A set of variables are correlated if they are neighbors of the same function node, e.g., y_1, y_2, y_3 with $f_1(y_1, y_2, y_3)$. We can see that two hidden variables are correlated iff their corresponding author nodes are linked by an edge on the candidate graph H', which means there is a potential advising relationship between them. And once a variable y_i changes its value, it will affect the value of all the functions corresponding to the potential advisor and advisee sets $Y_i \cup Y_i^{-1}$.

There is additional information stored in each variable node, as shown in the tables in Figure 5.2. y_i can take different values from Y_i, and the corresponding st_i, ed_i, and l_{ij} are pre-computed in stage 1.

Theoretically, one can incorporate any types of features into the TPFG model. For different kinds of relationships, the constraint can vary according to primary assumptions.

5.1.5   MODEL INFERENCE

To maximize the objective function and compute the ranking score along with each edge in the candidate graph G', we need to infer the marginal maximal joint probability on TPFG, according to Eq. (5.10). We first introduce the algorithm for general factor graph, discuss its deficiency, and then introduce a more efficient algorithm.

Sum-product + junction tree. There is a general algorithm called *sum-product* [Kschischang et al., 2001] to compute marginal function on a factor graph based on message passing. It performs exact inference on a factor graph without cycles. In the sum-product algorithm, the marginal functions of a single variable, a.k.a., messages, are passed between neighboring variable node and function node. Message passing is initiated at the leaves. The process terminates when two messages have passed on every edge. At each variable node, the product of all incoming messages is its marginal function. To compute the marginal maximal probability, we need to change sum-product to max-sum with a logarithmic transformation of the function value. If TPFG is a tree-structured factor graph, the message passing rule will be:

$$m_{y_i \to f_j 0}(y_i) = \sum_{j' \in Y_i \cup \{i\}, j' \neq j} m_{f_{j'} 0 \to y_i}(y_i) \tag{5.11}$$

$$m_{f_j 0 \to y_i}(y_i) = \max_{\sim \{y_i\}} (\log f_j(y_i, \{y_{i'}\}) + \sum_{i' \in Y_j^{-1} \cup \{j\}, i' \neq i} m_{y_{i'} \to f_j 0}(y_{i'})), \tag{5.12}$$

where $j' \in Y_i \cup \{i\}$, $j' \neq j$ represents $f_{j'}()$ is a neighbor node of variable y_i on the factor graph except factor $f_j()$, $\sim \{y_i\}$ represents all variables in $Y = \{y_1, \ldots, y_{n_a}\}$ except y_i.

Unfortunately, TPFG contains cycles. This algorithm cannot be applied directly. One solution to generalize it is a procedure known as *junction tree algorithm* Bishop [2006] for exact inference. The junction tree is a tree-structured undirected graph generated from arbitrary triangulated dependency graph, and can be solved by sum-product. Nevertheless, the computational cost of the algorithm is determined by the number of variables in the largest clique and will grow exponentially with this number in the case of discrete variables. The process to construct a junction tree alone consumes a lot in both time and space. In practice we found it fails to finish for 6000 variables, not to mention our TPFG has the scale of more than 600,000 variables.

To reduce the computational cost, we can do approximate inference instead of exact inference. A general method *loopy belief propagation* (LBP) Frey [1998] simply applies the sum-product algorithm in a cycle-containing graph. It passes message iteratively with flooding schedule. To

avoid repetitive information flow for multiple times through the graph, we design a special message passing schedule and the following algorithm according to the special property of TPFG.
New TPFG inference algorithm. The original sum-product or max-sum algorithm meet with difficulty since it requires that each node needs to wait for all-but-one message to arrive. Thus, in TPFG some nodes will be waiting forever due to the existence of cycles. To overcome this problem, we arrange the message passing in a mode based on the strict order determined by G'. Each node a_i has a descendant set Y_i^{-1} and an ascendant set Y_i.

The message is passed in a two-phase schema. In the first phase, messages are passed from advisees to possible advisors, and in the second, messages are passed back from advisors to possible advisees. Formally, there are two kinds of messages in the first phase: $m_{f_i() \to y_i}, m_{y_i \to f_j()}$ where $j \in Y_i$. The message from $f_i()$ to y_i is generated and sent only when all the messages from its descendants have arrived. And y_i immediately send it to all its ascendants $f_j(), j \in Y_i$. In phase two, there are also two kinds of messages: $m_{y_i \to f_i()}, m_{f_j() \to y_i}, j \in Y_i$, each of which are along the reverse direction on the edge as in phase 1. The messages are calculated as follows, derived from Eqs. (5.12) and (5.11):

$$m_{f_i() \to y_i}(x) = \max_{st_{ki} > ed_{ix}, \forall y_k = i} \left(\log l_{ix} + \sum_{k' \in Y_i^{-1}} m_{y_{k'} \to f_i()}(y_{k'}) \right) \qquad (5.13)$$

$$m_{y_i \to f_j()}(x) = m_{f_i() \to y_i}(x) \qquad (5.14)$$

$$m_{y_i \to f_i()}(x) = \sum_{j \in Y_i} m_{f_j() \to y_i}(x) \qquad (5.15)$$

$$m_{f_j() \to y_i}(x) = \max_{st_{kj} > ed_{jy_j}, \forall y_k = j} (\log l_{jy_j}$$

$$+ m_{y_j \to f_j()}(y_j) + \sum_{k' \in Y_j^{-1}, k' \neq i} m_{y_{k'} \to f_j()}(y_{k'})). \qquad (5.16)$$

After the two phases of message propagation, we can collect the two messages on any edge and obtain the marginal function.

$$r_{ij} = \max P(y_1, \dots, y_{na} | y_i = j)$$
$$= \exp \left(m_{f_i() \to y_i}(j) + m_{y_i \to f_i()}(j) \right). \qquad (5.17)$$

This algorithm still has redundant storage and computation. The messages sent between function nodes and variables nodes are function values, which need to be stored as vectors. Some messages are never used during the final merge, and some messages are simply transmitted from one variable node to its corresponding function node. We further simplify the message propagation by eliminating the function nodes and the internal messages between a function node and a variable node, and we find it equivalent to a message passing procedure on the homogeneous graph G', i.e., message propagation between authors, and the messages can be stored with each author in two vectors: one sent and one received. The order of messages passed is illustrated by the number on each edge in Figure 5.3. In this way both time and space are saved.

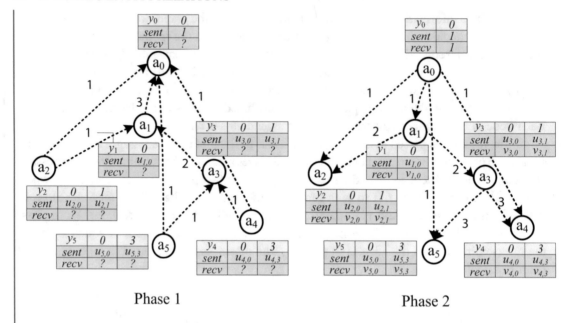

Figure 5.3: The two-phase message passing schema.

The improved message propagation is still separated into two phases. In the first phase, the messages **sent**$_i$ which passed from one to their ascendants are generated in a similar order as before. In the second, messages returned from ascendants **recv**$_i$ are stored in each node. After the two phases, each node collects the two vectors to generate the final ranking score. The derived rules are as follows:

$$\mathbf{sent}_{ij} = \log l_{ij} + \sum_{k \in Y_i^{-1}} \max_{st_{kx} > ed_{ij} \text{ or } x \neq i} \mathbf{sent}_{kx} \tag{5.18}$$

$$\mathbf{recv}_{ij} = \max_{j' \in Y_j, ed_{jj'} < st_{ij}} \left(\mathbf{recv}_{jj'} + \log l_{jj'} + \sum_{k \in Y_j^{-1}, k \neq i} \max_{x \in Y_k, st_{kx} > ed_{jj'} \text{ or } x \neq j} \mathbf{sent}_{kx} \right)$$

$$+ \sum_{x \in Y_i, x \neq j} \max_{j' \in Y_x} \left(\mathbf{recv}_{xj'} + \sum_{k \in Y_x^{-1}, k \neq i} \max_{x' \in Y_k, st_{kx'} > ed_{xj'} \text{ or } x' \neq x} \mathbf{sent}_{kx'} \right) \tag{5.19}$$

$$r_{ij} = \exp(\mathbf{sent}_{ij} + \mathbf{recv}_{ij}). \tag{5.20}$$

In the new algorithm, the message propagation can be done by using a stack-queue. In phase 1, each node will enter the queue once and the vector **sent**$_i$ for them is computed one by one. In phase 2, we scan the queue from the tail back to the head, i.e., treat it as a stack, and compute **recv**$_i$. Then we can normalize the results and collect them to get the ranking score. By using $\mathcal{O}(|\mathcal{E}_s'|)$ space, the running time of the algorithm can be reduced to $\mathcal{O}(\sum_{i=1}^{n_a} \delta_i \delta_i')$, where δ_i and

δ_i' are the in-degree and out-degree of each node a_i on graph G', respectively. As long as G' is sufficiently sparse, the maximal degree of the node can be seen as constant C and the complexity is further reduced to $\mathcal{O}(n_a)$.

5.1.6   EMPIRICAL ANALYSIS

Datasets. We use the DBLP computer science bibliography database. It consists of 654,628 authors and 1,076,946 publications with time provided (from 1970–2008). Ground truth is obtained from three sources: one is manually labeled by looking into the home page of the advisors, and the other two are crawled from the Mathematics Genealogy project[1] and AI Genealogy project.[2] We refer to them as MAN, MathGP, and AIGP, respectively. They only partially cover the authors in DBLP. We further separate MAN into three sub data sets: Teacher, PhD, and Colleague. Teacher contains all kinds of advisor-advisee pairs, while PhD only contains graduated PhDs pairing with their advisors. Colleague contains colleague pairs which are negative samples for advisor-advisee relationship. And we use these data to generate random data sets for test. See Table 5.1 for details.

Table 5.1:   Accuracy of prediction by P@$(2,\theta)$: $\frac{T}{T+F}$

data set	RULE	SVM	IndMAX		TPFG	
TEST1	69.9%	73.4%	75.2%	78.9%	80.2%	84.4%
TEST2	69.8%	74.6%	74.6%	79.0%	81.5%	84.3%
TEST3	80.6%	86.7%	83.1%	90.9%	88.8%	91.3%
TRAIN1=Colleague(491)+PHD(100)						
TEST1=Teacher(257)+MathGP(1909)+Colleague(2166)						
TRAIN2=TRAIN3=Teacher(257)+Colleague(2166)						
TEST2=PHD(100)+MathGP(1909)+Colleague(4351)						
TEST3=AIGP(666)+Colleague(459)						
IndMAX,TPFG: left - θ = 3rd quartile of $\{r_{ij}\}$; right - trained						

Method. We compare the proposed TPFG with the following baseline methods:

- Sum-Product+Junction Tree (JuncT). It computes the exact joint probability as the ranking score.

- Loopy Belief Propagation (LBP). It employs an approximate algorithm for inference.

- Independent Maxima (IndMAX). It computes the maximal local likelihood for each variable independently.

- SVM. It is a supervised approach and requires labeled pairs, both positive and negative, as training data.

[1]http://www.genealogy.math.ndsu.nodak.edu/
[2]http://aigp.eecs.umich.edu/

- RULE. For each author, from all the collaborators that satisfy Assumption 5.2, choose the one with most coauthored papers.

Effect of Network Structure

Using DFS with a bounded maximal depth d from the given set of nodes, denoted as DFS-d, we can obtain closures with controlled depth for a given set of authors to test. When d increases, the subnetwork grows larger until it is already the complete closure, i.e., the maximal connected subgraph of H containing the given set. We run TPFG on these closures and plot the ROC curves.

From Figure 5.4(a) we see that for closures with different depths, TPFG achieves better accuracy when the depth increases, and they all outperform the IndMAX method by more than 5% in AOC. And on the complete closure TPFG reaches the same accuracy as on the whole network since disconnected components will not affect each other.

On these various scaled subnetworks, TPFG achieves different level of approximations to the optimal global joint probability on the whole network. To compare it with the exact maximal joint probability and other approximate algorithm, we show the result on a small graph due to limitations of JuncT and LBP (see Section 5.1.5). The small graph is constructed by extracting the nodes in PhD and their advisors, and then building 1-closure of it. It consists of 1310 nodes. From Figure 5.4(b) we find that in the small graph TPFG approximates well to the exact inference algorithm JuncT(AOC difference < 0.01), and outperforms LBP by 16.9%.

Effect of Training Data

Support Vector Machines (SVMs) are accurate supervised learning approaches and shown to be successful in syntax-based relation mining [Coppola et al., 2008]. If we treat advisor-advisee pairs as positive examples and non advisor-advisee pairs as negative examples, we can reduce advisor mining to a classification problem on the ordered pairs (a_i, a_j). In this setting it requires to define some features for each pair of coauthors, and train the classifier by feeding both positive and negative samples. For fair comparison, we combine Kulczynski and IR measures with what were used in Yang et al. [2009] as features.

Direct application of SVM only shows whether a given pair is an advisor-advisee pair, and it is often the case an author is predicted to have multiple advisors. Thus, we examine the probabilistic scores in the test data, and rank them to draw the ROC curve. TPFG is 4.2% and 2.4% higher in AOC than SVM in TEST1 and TEST2, respectively.

TPFG can utilize the training data in a simple way. The parameter θ in $P@k, \theta$ can be optimized according to certain criteria such as achieving best information gain on the training data. Then we use the trained parameters to do predictions on test data. Table 5.1 shows the improvement by utilizing the training data. After this simple training, TPFG can reach an accuracy of 84–91%.

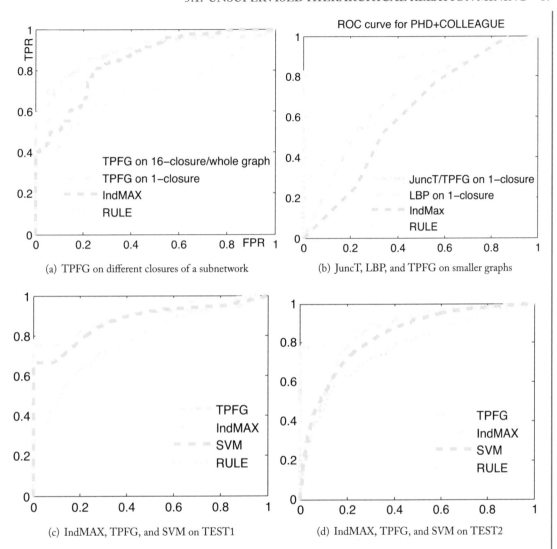

Figure 5.4: ROC curves for advisor-advisee prediction.

SVM actually makes a supervised combination of all the assumptions and rules used in TPFG. The difference lies in that it does not explore the constraint and dependency relying on the whole network structure. It does a fairly good job, but still 5–10% worse than optimized TPFG. In conclusion, TPFG can achieve comparable or even better accuracy compared with a supervised method. When parameters are adjusted with training data, its accuracy can be further improved by around 3%.

Case Study

With case study, we find that TPFG can discover some interesting relations beyond the "ground truth" from single source. Table 5.2 shows some examples. The ranking results provided with advising time facilitate finding such kind of advising relations, which cannot be easily discovered by referring to genealogy projects. The mean of deviation of estimated graduation time to the labeled time on the test data sets is 1.76–1.78.

At least 40% of the error is contributed by name ambiguity. For example, if we try to find the advisor for "Joseph Hellerstein," the algorithm returns wrong results. If we distinguish "Joseph M. Hellerstein" and his publications properly, our algorithm is able to find the 'half' right answer Michael Stonebraker ranked top 1. The answer is half right because there is a co-advisor Jeffrey F. Naughton, who is also ranked high in top 15%. Duplication are even more common among Chinese names. Other reasons for false negatives include that one researcher collaborated with multiple advisors or that one coauthored fewer papers with the advisor. The latter case happens more often for older researchers, for whom the publication data are not as complete as nowadays. In those situations, it is almost impossible to find credible advisor-advisee relationships merely based on their publication records. The inference algorithm can find typical cases but will miss such atypical cases.

Table 5.2:   Examples of mined relations. Time—the estimated advising time; Note—the factual relation and graduation year

Advisee	Top Ranked Advisor	Time	Note
David M. Blei	1. Michael I. Jordan	2001-2003	PhD advisor, 2004
	2. John D. Lafferty	2005-2006	Postdoc, 2006
Hong Cheng	1. Qiang Yang	2002-2003	MS advisor, 2003
	2. Jiawei Han	2004-2008	PhD advsior, 2008
Sergey Brin	1. Rajeev Motwani	1997-1998	'Unofficial advisor'[3]

[3] cited from a blog of Sergey Brin, who left Stanford to found Google around 1998.

Scalability Performance

TPFG is shown to be scalable in Figure 5.5. With regard to the inference time, TPFG costs only 13 s on the whole DBLP dataset. As a classification approach, SVM's scalability is related to the size of training data. As an example, the feature computation takes one hour and a half, and the model inference takes 31 s for Train1 and 6 min 26 s for Train2.

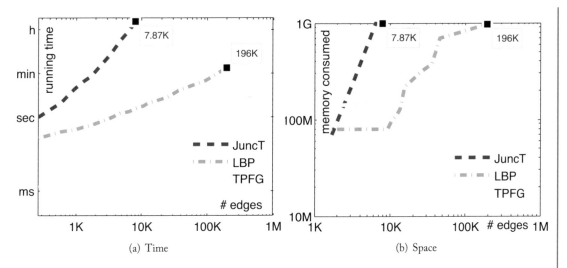

(a) Time   (b) Space

Figure 5.5: Scalability results in log-log scale. JuncT and LBP fail at 10K and 200K, respectively, due to memory limitation.

5.2 SUPERVISED HIERARCHICAL RELATION MINING

In the supervised setting, we have more complex features, constraints, and knowledge propagation rules with unknown importance. We need to learn their importance from the data. Also, we need to find a generic model to handle different types of signals.

Example 5.1 Given a set of named people and with knowledge of their ages, nationality and mentions from text, we want to find the *family* relation among them. Besides text features such as the coreferential link between two names with family keywords, we have constraints like: *(1) for two persons to be siblings, they should have the same parent; and (2) if A is B's parent, then A is unlikely to be C's parent if B and C have different nationalities.* The output should be a family tree or forest of these people.

Example 5.2 In an online forum, we want to predict the replying relationship among the posts within the same thread, with knowledge of the content, author and the posting time. Every post replies to one earlier post in the same thread, except for the first one. One intuition is that *a post will have similar content with the one it replies to*; yet another possibility is *two similar posts may reply to a common post*. The output is a tree structure of each threaded discussion.

To concretely explain this approach, we focus on the problem of mining hierarchical relations. However, our methodology can be applied to general relation mining problems as well.

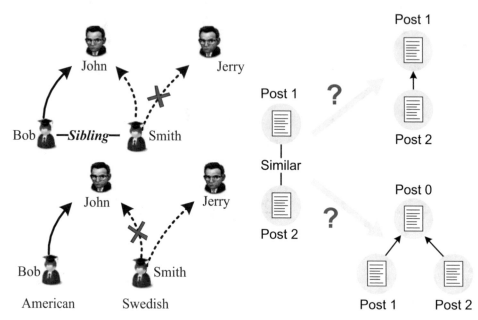

(a) Two soft dependency rules on family tree: the relative importance need be learnt

(b) Conflict rules on forum reply structure: similar posts may be filiation or siblings

Figure 5.6: Examples for the dependency among the relations.

The goal is to find a hierarchy, in which each node represents an object, and each edge represents a certain relation between the two objects, such as advisor-advisee.

Definition 5.3   Given a set of objects $V = \{v_1, \ldots, v_n\}$ and their directed links $\mathcal{E} \in V \times V$, a relation $R \subset \mathcal{E}$ is a **hierarchical relation** if (1) for every object $u \in V$, there exists exactly one object $v \in V$ such that $(u, v) \in R$; and (2) there does not exist a cycle along R, i.e., no sequence $(u_1, \ldots, u_t), t > 1$, such that $u_1 R u_2, \ldots, u_{t-1} R u_t, u_t R u_1$. In the relation instance $v_i R v_j$, we name v_i as a *child* and v_j a *parent*. For convenience we denote (v_i, v_j) by e_{ij}. We define the out-neighbors of one node as $Y_i = \{v_j | (v_i, v_j) \in \mathcal{E}\}$.

In general, we want to predict for every pair of directly linked objects $(v_i, v_j) \in \mathcal{E}$, whether the statement '$(v_i, v_j) \in R$' is true.

Figure 5.7 gives one example. If we instantiate it as a family tree prediction problem, each node v_i represents a person, and each link points from one person to his potential parent. v_1 and v_2 each has a link to themselves, implying they can be the root of a tree. We call $G = (V, E)$ a *candidate graph*. Suppose we know the ages of these people, we can make sure the candidate graph has no directed cycles by always linking a younger person to an older one. Many other real problems also have this property.

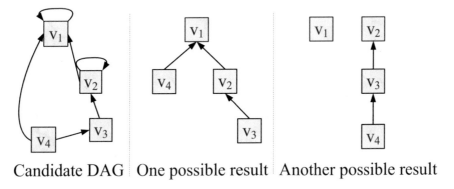

Figure 5.7: An illustration of the problem definition.

Assumption 5.3 *The candidate graph is a directed acyclic graph (DAG).*

With this assumption, given a set of objects, a tree-like structure can be learned in three steps. First, we extract a partial order for the objects and build a DAG. Next, we learn a model with labeled relations, and finally conduct prediction for unknown relations.

We introduce a generic discriminative model CRF-Hier and use the above examples as cases for study.

5.2.1 CONDITIONAL RANDOM FIELD FOR HIERARCHICAL RELATIONSHIP

We model the joint probability of every possible relationship $(v_i, v_j) \in \mathcal{E}$ being a truly existent relationship in R. We use an indicator variable x_{ij} for the event "$(v_i, v_j) \in R$," i.e., $x_{ij} = 1$ if $(v_i, v_j) \in R$, and 0 if not. As we have analyzed, the inference of the relationship for some pairs are not independent. Suppose we have evidence that two people v_2 and v_4 are not siblings, we may not expect the two events "$(v_2, v_1) \in R$" and "$(v_4, v_1) \in R$" to happen together. We formalize that intuition as a Markov assumption that events involving common objects are dependent.

Assumption 5.4 *Two events "$(v_a, v_b) \in R$" and "$(v_c, v_d) \in R$" correspond to connected random variables in a Markov network if and only if they share a common node, i.e., one of the following is true: $v_a = v_c, v_b = v_c, v_a = v_d$ or $v_b = v_d$. (Figure 5.8)*

It immediately follows that the Markov network can be derived from the candidate graph G by having a node for every edge in G and connecting nodes that represent two adjacent edges in G, namely the *line graph* of G.

The conditional joint probability is formulated as

$$p(X|G, \Theta) \propto \exp\left(\sum_{k=1}^{K} \theta_k F_k(X, G) + C(X, G)\right), \tag{5.21}$$

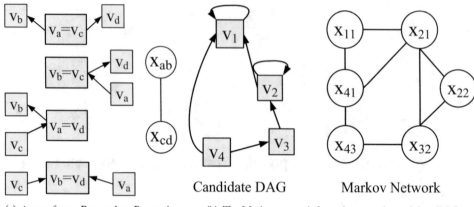

(a) 4 cases for $v_a R v_b$ and $v_c R v_d$ to be directly dependent

(b) The Markov network from the example candidate DAG

Figure 5.8: Illustration of the Markov assumption.

where $\{F_k(X, G)\}_{k=1}^{K}$ is a set of features defined on the given candidate graph G and the indicator variables $X = \{x_{ij}\}$; $\{\theta_k\}_{k=1}^{K}$ are the weights of the corresponding features. C is a special feature function to enforce the hierarchy constraint $\sum_{(v_i, v_j) \in \mathcal{E}} x_{ij} \leq 1$:

$$C(X, G) = \begin{cases} -\infty & \exists i, \sum_{(v_i, v_j) \in \mathcal{E}} x_{ij} > 1 \\ 0 & o.w. \end{cases} \quad (5.22)$$

Any other hard constraints can be encoded in the same manner.

Thus, once we learn the weights $\{\theta_k\}_{k=1}^{K}$ from training data, the relation inference task could be formulated as a maximum a posterior (MAP) inference problem: for each given candidate DAG G, we target for the optimal configuration of the relationship indicator X^*, such that,

$$X^* = \arg \max_{X \in \mathcal{X}} p(X|G, \Theta), \quad (5.23)$$

where $\mathcal{X}$ is the set of all the possible configurations, i.e., the search space. Since every x_{ij} can take 0 or 1, the size of the space is $2^{|\mathcal{E}|}$.

Such a formulation keeps the maximal generality, while it poses great challenge to solve the combinatorial optimization problem. We can improve it in two ways.

First, we can explore the form of feature functions F_k. Each of them can be defined on all variables, and then the computation of every function value relies on the enumeration of X in $\mathcal{X}$. However, since we have made the Markov assumptions, the dependency can be represented by a factor graph, and each feature function can be decomposed into the potential functions over cliques of the graph (Hammersley-Clifford theorem). Moreover, we can restrict the potential functions to have interactions between at most a pair of random variables x_{ij} and x_{st}. In fact, we have the following claim: any factor graph over discrete variables can be converted to a factor graph restricting to pairwise interactions, by introducing auxiliary random variables. This property is found by Weiss and Freeman [2001]. Although the generic procedure of conversion may introduce additional variables, we will show that quite a broad range of features can be materialized in as simple forms as pairwise potentials without the help of auxiliary variables in the next section. Here we factorize the constraint H to exemplify the philosophy: $C(X, G) = \prod_{e_{is},e_{it} \in \mathcal{E}} h(x_{is}, x_{it})$, where $h(x_{is}, x_{it}) = -\infty$ if $x_{is} = x_{it} = 1$, and 0 otherwise.

Next, we can reduce the number of variables and constraints. To leverage the constraint that one node has at only one parent and the assumption that the candidate graph G is a DAG, we introduce a variable y_i to represent the parent of v_i, i.e., $y_i = j$ if and only if $e_{ij} = (v_i, v_j)$ is an instance in a hierarchical relationship R. Given the DAG, the problem is equivalent to the task of predicting y_i's value from Y_i.

Hypothesis 5.4 implies the existence of two kinds of dependencies: two variables y_i and y_j where v_j is a candidate parent of v_i; two variables y_s and y_t such that v_s and v_t share a common candidate parent v_m. With this formulation, the constraint C is not needed any more, and the objective function has the following form:

$$p(Y|G, \Theta) \propto \exp\left(\sum_{k \in I_1} \sum_{y_s \in S'_k} \theta_k g_k(y_s|Y_s) + \sum_{k \in I_2} \sum_{(y_i,y_j) \in P'_k} \theta_k f_k(y_i, y_j|Y_i, Y_j) \right), (5.24)$$

where I_1 and I_2 are the index sets of features that can be decomposed into singleton potential and pairwise potential functions respectively, and S_k and P_k are the decomposed singleton and pairwise cliques for the k-th feature.

As an example, for the candidate graph G in Figure 5.9 we will build a factor graph like the one on its right. Four nodes v_1, v_2, v_3, and v_4 have four parent variables y_1 to y_4, while the range for them is $Y_1 = \{v_1\}$, $Y_2 = \{v_1, v_2\}$, $Y_3 = \{v_3\}$, and $Y_4 = \{v_1, v_3\}$. For each variable there can be one or more singleton potential functions. For each pair of directly linked nodes, we have a pairwise potential function f_4 in this example (the one between y_1 and y_4 is omitted). v_2 and v_4 have a common parent candidate, so there is one pairwise potential f_5 defined on y_2 and y_4.

We name this model CRF-Hier as it is a conditional random field optimally designed for the hierarchical relationship learning problem.

Let us look at several examples for specific potential definitions.

Example 5.4   The names of a parent and a child should appear in the same sentence (support sentence) together with family relationship words like "son, father."

$g(y_i|Y_i) = \#$support sentences of (v_i, v_{y_i}).

Example 5.5   Siblings have common parent.

$f(y_i, y_j|Y_i, Y_j) = [y_i = y_j]p(v_i \text{ and } v_j \text{ are siblings})$. where we use Iverson bracket to denote a number that is 1 if the condition in square brackets is satisfied, and 0 otherwise. In this case, $[y_i = y_j] = 1$ if $y_i = y_j$ and 0 otherwise.

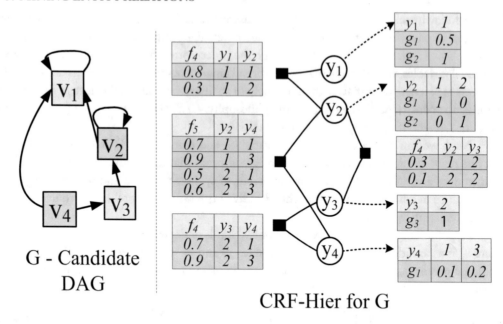

Figure 5.9: The graphical representation of the proposed CRF-Hier model for the example in Figure 5.7. All the singleton potentials defined on each variable are listed in the table connected to the related variables by dotted line; pairwise potentials are represented by solid rectangles and tables listing the pairwise function values when the pair of variables take different configurations.

Example 5.6 Two people of the same age are unlikely siblings.

$$f(y_i, y_j | Y_i, Y_j) = [y_i = y_j][v_i\text{'s age} = v_j\text{'s age}].$$

Note that these statements are not always true, e.g., twins can be siblings. But the advantage of the probabilistic framework is that it does not enforce them to be true. The indicator in Example 5.6 takes value 0 or 1, and the weight of the corresponding feature controls how much we trust this rule. Eventually the compatibility with this rule plays a factor as the product of potential function and feature weight in the additive log-likelihood.

We see that these different features, constraints and propagation rules can be encoded in a unified form; we just need to define the potential for each of them. We discuss how to systematically design potentials in the next subsection, followed by the inference and learning algorithm.

5.2.2 POTENTIAL FUNCTION DESIGN

We have restricted our focus to the features that can be decomposed into either singleton potential or pairwise potential functions. We summarize the potential types with domain-independent cognitive meanings so that one can design domain-specific potentials.

Let us begin with important singleton potentials.

- *Homophyly.* The first kind, and probably most widely applicable, is a similarity measure between two objects. The assumption here is that the filiation connects to homophily, e.g., content similarity, interaction intensity (e.g., the frequency of telephone calls in unit time), location adjacency, time proximity (e.g., whether two documents are published within a short period). This kind of similarity measure sim is symmetric, and there are numerous metrics we can use. The potential function has the form $g(y_i|Y_i) = sim(v_i, v_{y_i})$.

- *Polarity.* The second kind, almost equally important, is an asymmetric similarity measure. It is used to measure the dominance of certain attributes of the parent on the child, e.g., authority difference, bias of interaction tendency (e.g., whether A writes many more emails to B than B to A), the degree of conceptual generalization/specialization. Such measure $asim$ quantifies the partial order in terms of polarity between linked nodes, in the form of $g(y_i|Y_i) = asim(v_i \rightarrow v_{y_i})$.

- *Support pattern.* The third kind of potentials characterize the preference to certain patterns involving a pair of nodes with filiation. We can define a potential based on the number of pattern occurrences: $g(y_i|Y_i) = |SP(v_i, v_{y_i})|$, where SP denotes the support pattern set.

The pairwise potentials are responsible for the knowledge propagation as well as the restrictive dependencies.

- *Attribute augmentation.* One intuition for the knowledge propagation is that one node can inherit attributes from its parent or child to augment its own. In our model this can be realized by defining a pairwise potential $f(y_i, y_j|Y_i, Y_j) = [y_i = j]sim(v_i, v_{y_j})$. It can be elaborated in two ways: knowing that the parent of v_i is v_j, v_j's parent will tend to choose a similar one with v_i; or given that the parent of v_j is v_{y_j}, it affects the decision of its child towards inheriting attributes from its parent. By replacing the boolean indicator $[y_i = j]$ with a weighting function of v_i and v_j we can control the extent to which we propagate the attribute.

- *Label propagation.* Sometimes we can measure how likely two nodes share the same parent, so the label of one's parent can be propagated to similar nodes with a function like $f(y_i, y_j|Y_i, Y_j) = [y_i = y_j]sim(v_i, v_j)$. Given v_i's parent v_{y_i}, the more similar v_j is with v_i, the larger contribution to the joint likelihood this function will have via setting v_j's parent to be the same, i.e., $y_j = y_i$.

- *Reciprocity.* This kind of potentials can handle a more complicated pattern that occurs in child-parent and parent-child pairs alternatively. For example, "if author A often replies author B, then author B is more likely to reply author A." For such kind of rule, we seem to need a big factor function like $F(y_i|G, Y \setminus y_i) = \sum_{A,B}([a_i = B][a_{y_i} = A] \sum_{a_j=A}[a_{y_j} = B])$, where a_i stands for v_i's author. It requires all labels to be known. Fortunately, we find that rules in this form have equivalent decomposed representation. The above rule can be decomposed into pairwise potentials $f(y_i, y_j|Y_i, Y_j) = [a_i = a_{y_j}][a_{y_i} = a_j]$.

Table 5.3: Potential categorization and illustration

Type	Cognitive Description	Potential Definition	Example
homophyly	parent and child are similar	$g(y_i) = sim(v_i, v_{y_i})$	textual similarity, interaction intensity, spatial and temporal locality
polarity	parent is superior to child	$g(y_i) = asim(v_i \rightarrow v_{y_i})$	authority, interaction tendency, conceptual extension and inclusion
support pattern	patterns frequently occurring with child-parent pairs	$g(y_i) = |SP(v_i, v_{y_i})|$	contextual pattern, interaction pattern, preference to certain attributes
forbidden pattern	patterns rarely occurring with child-parent pairs	$g(y_i) = -|FP(v_i, v_{y_i})|$	forbidden attribute to share, forbidden distinction
attribute augment	use inherited attributes from parents or children	$f(y_i, y_j) = [y_i = j]sim(v_i, v_{y_j})$	content propagation for documents, authority propagation for entities
label propagate	similar nodes share similar parents (children)	$f(y_i, y_j) = sim(v_{y_i}, v_{y_j})sim(v_i, v_j)$	siblings share common parents, colleagues share similar supervisors
reciprocity	patterns altering in child-parent and parent-child pairs	$f(y_i, y_j) = sim(v_i, v_{y_j})sim(v_j, v_{y_i})$	author reciprocity in online conversation – forth-and-back;
constraints	restrict certain patterns	$f(y_i, y_j) = -|CP(v_i, v_j, v_{y_i}, v_{y_j})|$	consistency of transitive property

For conciseness we omit the conditional variable in g and f.
SP = support pattern set, FP = forbidden pattern set, CP = constraint pattern set.

For a specific application we can encode an arbitrary number of potentials of each type. In principle, one can apply frequent pattern mining or statistical methods to find distinguishing features in each category. When multiple features are used, the model needs to be trained to reach a compromise.

5.2.3 MODEL INFERENCE AND LEARNING

Given a training set of network $G = \{\mathcal{V}, \mathcal{E}\}$ with labeled instances with hierarchical relationship L, we need to find the optimal model setting $\Theta = \{\theta_k\}_{k=1}^K$, which maximizes the conditional likelihood defined in Eq. (5.24) over the training set. Let $Y^{(o)}$ and T be the variable sets and assignments corresponding to labeled relation instances L, and $Y^{(u)}$ be the unknown labels in training data. Their union is the full variable set Y in training data. The log-likelihood of the labeled variables could be written as:

$$
\begin{aligned}
L_\Theta &= \log p(Y^{(o)} = T | G, \Theta) \\
&= \log \sum_{Y^{(u)}} \exp\left(\Theta^t \mathbf{F}(Y|_{Y^{(o)}=L}, G)\right) - \log Z(G, \Theta),
\end{aligned} \tag{5.25}
$$

where $\mathbf{F}(Y, G)$ is the vector representation of the feature functions, $Z(G, \Theta) = \sum_Y \exp(\Theta^t \mathbf{F}(Y, G))$.

To avoid overfitting, we penalize the likelihood by L2 regularization. Taking the derivative of this object function, we could get:

$$
\nabla L_\Theta = E_{p(Y^{(u)}|G,\Theta,Y^{(o)}=T)}\mathbf{F}(Y|_{Y^{(o)}=T}, G) - E_{p(Y|G,\Theta)}\mathbf{F}(Y, G) - \lambda\Theta. \tag{5.26}
$$

When the training data are fully labeled, $Y^{(u)} = \emptyset$, and Eq. (5.26) becomes simply:

$$
\nabla L_\Theta = \mathbf{F}(Y = T, G) - E_{p(Y|G,\Theta)}\mathbf{F}(Y, G) - \lambda\Theta.
$$

The first part is the empirical feature value in the training set, the second part is the expectation of the feature value in the given training data, and the last part is the L2-regularization. Given that the expectation of the feature value can be computed efficiently, L-BFGS [Byrd et al., 1994] algorithm can be employed to optimize the objective function Eq. (5.25) by the gradient. Although the objective is not convex with respect to Θ, we can expect to find a local maximum of it. When there are multiple training networks, we train the model by optimizing the sum of their log likelihood.

When the feature weights are fixed, the learning process requires an efficient inference algorithm for marginal probability of every clique: singletons and edges where we define the potentials. The prediction, however, requires MAP inference to find maximal joint probability.

In the toy model in Figure 5.9, there are 5 different potentials g_1, g_2, g_3, f_4, f_5. The algorithm will learn the weight vector $\Theta = \{\theta_1, \ldots, \theta_5\}$ from training data, and then find the optimal value of $Y = \{y_1, \ldots, y_4\}$ to predict the structure.

5.2.4 EMPIRICAL ANALYSIS

Evaluation Measure

Few studies have been carried out on how we should evaluate the quality of hierarchical relationship prediction. Accuracy on the predicted parent y_i (A_{par}), and the accuracy on the predicted relation pairs x_{ij} (A_{pair}), are two most natural evaluation criteria; they or their variants (Precision, Recall, etc.) are typically employed. However, such measures only evaluate the prediction variables on each node or each edge in an isolated view, missing some aspects of the comprehensive goodness of structure. We take an example to illustrate. Figure 5.10 lists a ground-truth structure and several different reconstruction results. Both (b) and (c) have the same A_{par} and A_{pair} because only one node has incorrect predicted parent. However, the chain is quite different from the gold

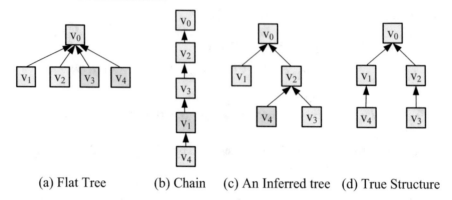

(a) Flat Tree   (b) Chain   (c) An Inferred tree   (d) True Structure

Figure 5.10: Comparison of inferred structures and the gold standard structure. (a)–(c) are three possible prediction results for the true structure in (d). Green nodes have right prediction for their parents, while red nodes have wrong predictions .

standard tree with two branches, and result (c) should be regarded closer to the ground truth. So we can see that one mistake may not only affect the parent of one node, but also deviate the shaping statistics of other nodes (e.g., in their degrees, number of ancestors, and descendants). In other words, different edges have different importance in preserving the shape of the tree, which is not reflected by the unweighted judgment of each prediction. Tree similarity measures for ontologies such as tree edit distance [Bille, 2005] are neither desirable because the edit operations do not apply here, and the computation of them has biquadratic complexity w.r.t. the tree size.

   We define a set of novel measures for quantitative evaluation of the quality of hierarchical relationship prediction, which can be computed in linear time. Formally, let T be the ground-truth structure for n linked objects $V = \{v_1, \ldots, v_n\}$, and Y the prediction. We evaluate how well the structure is preserved in prediction by examining two additional aspects: the ancestors and the path to root.

- Precision and recall of ancestors P_{anc}, R_{anc}:

$$P_{anc} = \frac{\sum_{i=1}^{n} \delta[Anc_i(Y) \subset Anc_i(T)]}{n} \tag{5.27}$$

$$R_{anc} = \frac{\sum_{i=1}^{n} \delta[Anc_i(T) \subset Anc_i(Y)]}{n}, \tag{5.28}$$

where $Anc_i(Y)$ and $Anc_i(T)$ stand for the ancestor set of node v_i in prediction and in ground truth, respectively.

- Accuracy of the path to root, A_{path}:

$$A_{path} = \frac{\sum_{i=1}^{n} \delta[path_i(Y) = path_i(T)]}{n}, \tag{5.29}$$

where $path_i(Y)$ and $path_i(T)$ is the path from node v_i to its root in prediction and in ground truth, respectively. A_{path} measures whether we can trace from each particular node to root without any mistake, thus it is the most strict measure.

As a commonly used compromise between precision and recall, we can also define F-value for the proposed measure for ancestors as the harmonic mean of precision and recall. For the example in Figure 5.10, the proposed metrics have the values as shown in Table 5.4. We can see that although (b) and (c) have the same accuracy on the parent prediction, three of the four proposed measures all imply the inferred structure in (c) has better quality than the chain. Also, we notice that A_{path} is a most strict measure, and in that measure (b) is even worse than (a). That implies one predicted structure may be good in some aspect but bad in others. This reaffirms the necessity of using multiple measures for the tree structure evaluation.

The hardness of the inference problem highly depends on the number of candidate parents of each node. Random guess will have much lower performance than 0.5 in most cases, even when each node only has two or three candidate parents.

Table 5.4: Measurement for structures in Figure 5.10

Structure	P_{anc}	R_{anc}	$F1_{anc}$	A_{path}	A_{par}
(a) flat tree	1.00	0.60	0.75	0.60	0.60
(b) chain	0.60	1.00	0.75	0.40	0.80
(c) inferred	0.80	0.80	0.80	0.80	0.80

Uncovering Family Tree Structure

We apply the method to an entity relation discovery problem. In the Natural Language Processing (NLP) community, it is sometimes studied as a slot filling task [Ji and Grishman, 2011], i.e., to answer questions like "who are the top employees of IBM." Some relation types satisfy the definition of hierarchical relationship, e.g., manager-subordinate, parent_organization-subsidiary, country-state-city. We take the family relation (parent-child) as the case to study, and try to answer the following two questions: (1) whether the proposed method works better than the state-of-the-art NLP approaches to generic entity relation mining; and (2) how good a joint model is compared to a model that does not handle dependency rules, or uses the rules just for post processing.

For clear demonstration, we define the task as automatically assembling the family tree from a set of named person entities. These named entities were extracted from two famous American families, *Kennedy family* and *Roosevelt family*. Given any pair of named entities, analysis of Web documents yields features such as co-occurrence statistics, residence, age, birth-date, and death-date. Table 5.5 lists the potentials for this task.

The results of random guess reflect the hardness of the problem. Three baselines are compared: (1) NLP, the general relation mining approaches based on NLP techniques described

Table 5.5: Potentials used in family tree construction task

Type	Potential Description
homophily	v_i and v_{y_i} live in the same location? mutual information of two names and family keywords from web snippets
polarity	suffix comparison: Junior, I, III, IV *etc.*
support pattern	co-occurrence of two names and some family keywords from web snippets; parent-child implied by Wikipedia infoboxes
forbidden pattern	child's birth year after parent's death+1; non parent-child implied by Wikipedia
attribute augment	same residence location of one and grandparent
label prop	people who are likely to be siblings share the same parent
constraints	people with the same age get lower possibility to share the same parent

by Chen et al. [2010]; (2) ranking SVM, a robust machine learning technique developed from support vector machine (SVM) for ranking problem, but only singleton features can be handled; and (3) ranking SVM + post processing (PP). For Ranking SVM, we treat each node as a query, and its parent as relevant "document" and all the other candidates as non-relevant "documents." For post processing, some pairwise potentials as global constraints (e.g., one person cannot have multiple parents; siblings should share the same parents) are encoded in Integer Linear Programming (ILP), in order to maximize the summation of confidence values from Ranking SVM subject to these constraints.

Table 5.6 shows the results on two families with 60 and 40 members, respectively. We can see that the optimized model for hierarchical relationship discovery performs 2- to 3-fold better in most measures than the general purpose relation miner. Compared with the two-stage method that uses post processing rules, the joint model can better integrate them into the learning and inference framework. The margin is the largest in the most strict measure path accuracy (333%, 133%). That implies CRF-Hier makes fewer mistakes at the key positions of the tree structure, where the chance of absorbing knowledge and doing regularization is higher. The post processing does not help much because the confidence estimation from the prediction is not always reliable due to noises in prediction features. As one example, the prediction component successfully identified "*William Emlen Roosevelt*" as the parent of "*Philip James Roosevelt*" but with a low confidence; while it mistakenly identified "*Theodore Roosevelt*" as the parent of "*George Emlen Roosevelt*" but with a much higher confidence due to their high mutual information value in Web snippets. Therefore, in the post-processing stage, based on the fact that *"Philip James Roosevelt"* and *"George Emlen Roosevelt"* are siblings, the label propagation rule mistakenly changed the parent of *"Philip James Roosevelt"* to *"Theodore Roosevelt."* In CRF-Hier model, the weights of the local features and the propagation rules are learned from the data, and the global optimization prevents this mistake.

Table 5.6: Prediction performance for family tree

Train/Test	Method	$F1_{anc}$	A_{path}	A_{par}
Train on Kennedy, test on Roosevelt	Random	< 0.01	< 0.01	0.0943
	NLP	0.1146	0.0333	0.1833
	RankingSVM	0.0667	0.0500	0.5000
	RankingSVM+PP	0.0667	0.0500	0.5833
	CRF-Hier	**0.3439**	**0.2167**	**0.7167**
Train on Roosevelt, test on Kennedy	Random	< 0.01	< 0.01	0.1313
	NLP	0.2625	0.1750	0.2500
	RankingSVM	0.4371	0.1500	0.3250
	RankingSVM+PP	0.4750	0.1500	0.3500
	CRF-Hier	**0.4846**	**0.3500**	**0.4000**

Uncovering Online Discussion Structure

Online conversation structure is a popular topic studied by researchers in information retrieval, since the structure benefits many tasks including keyword-based retrieval, expert finding, and question-answer discovery. We perform the study on finding reply relationship among posts in online forums. The data are crawled from Apple Discussion forum (http://discussions.apple.com/) and Google Earth Community (http://bbs.keyhole.com/). The posts in each thread can be organized as a tree based on their reply relationship. The task is to reconstruct the reply structure for the threads with no labels, given a few threaded posts with labeled reply relationship.

The features of each type are listed in Table 5.7. The competitor is Ranking SVM, used by Seo et al. [2009] for this task. Again, it can only handle singleton features. We also compare with a naive baseline that always predicts chain structure.

To see how many labeled data are needed to achieve good performance, we fix the test data of 2000 threads and vary the training data size in two different ways. First, we use all the labels from each thread, but vary the size of training threads from 50–2500. Second, we fix the number of training threads as 1000, and change the number of labels we use for each thread from 3–11. From Figure 5.11(a), we find that with a small training set, 50 labeled threads, CRF-Hier already achieves encouraging performance. The margin is significant because even the naive baseline of predicting every post to reply to the last post gives 0.74 in $F1_{anc}$ — compared with Ranking SVM's 0.80 and CRF-Hier's 0.86, CRF-Hier doubles the margin of what can be achieved by Ranking

Table 5.7: Potentials used in post reply structure task

Type	Features, Rules and Constraints
homophily	tf-idf cosine similarity $\cos(v_i, v_{y_i})$; recency of posting time
polarity	whether v_{y_i} is the first post of the thread; whether v_{y_i} is the last post before v_i
support pattern	whether a_{y_i}'s name appears in post v_i, where a_{y_i} is the author of v_{y_i}
forbidden pattern	an author does not reply to himself
attribute augment	the average content of one post's children is similar to its parent: $[y_i = j]\cos(v_i, v_{y_j})$
label propagate	similar posts reply to the same post: $[y_i = y_j]\cos(v_i, v_j)$
reciprocity	author A replies B, motivated by B replying A: $[a_{y_j} = a_i][a_{y_i} = a_j]$
constraints	one author does not repeat reply to a post; B replying A's post, A does not reply to B's earlier post: $-[t_{y_i} < t_j][a_{y_i} = a_j][a_{y_j} = a_i]$

Table 5.8: Cross domain evaluation (CRF-Hier/Ranking SVM)

$F1_{anc}$	Train \ Test	Apple	Google Earth
	Apple	**0.8476**/0.8136	**0.8233**/0.7855
	Google Earth	**0.8383**/0.8017	**0.8186**/0.8099
A_{path}	Train \ Test	Apple	Google Earth
	Apple	**0.7326**/0.6722	**0.6909**/0.6325
	Google Earth	**0.7143**/0.6548	**0.6797**/0.6610

All improvements are statistically significant with $p < 0.05$

SVM from the naive baseline. As more training data is added, the testing performance is relatively more stable than Ranking SVM. Figure 5.11(b) shows that when the labels for each tree is very incomplete, CRF-Hier degrades its performance to its competitor because the pairwise features cannot be well exploited. When the labels become reasonably sufficient (5 posts in this case) to characterize the structural dependency among them, CRF-Hier presents superiority (increase the margin from baseline by 32% in A_{path}). Although CRF-Hier has more feature weights to learn, the L2 regularization mitigates overfitting, and the model works well even when the training data size is small.

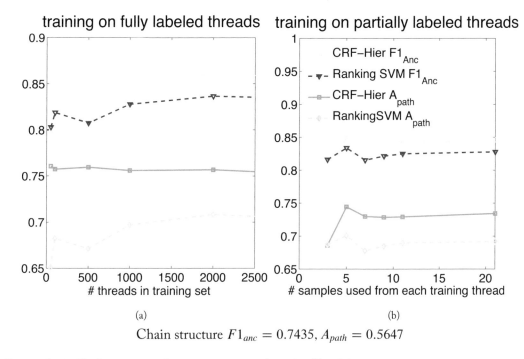

Chain structure $F1_{anc} = 0.7435$, $A_{path} = 0.5647$

Figure 5.11: Performance with varying training data size (Apple).

To see the adaptability of the model, we randomly select 2,000 threads for training and 2,000 threads for testing from each of the two datasets, and perform a cross domain experiment with the 4 combinations of train/test sets. Apple Discussion is a computer technical forums, while Google Earth focuses on entertainments. From the comparative results in Table 5.8, we can find that with the help of pairwise features, CRF-Hier generalizes better than Ranking SVM which relies on the singleton features only.

5.3   SEMI-SUPERVISED CO-PROFILING

In most social networking websites, a few users voluntarily provide their attributes in their online profiles (e.g., 40% of Facebook users provide their employers [Mislove et al., 2010], and 20% of Twitter users provide their home cities [Li et al., 2012]). In such cases, we can leverage the labeled links to certain relational attributes, as well as the unlabeled connections for relationship mining. In this section, we present a graph-based regularization framework for co-profiling in this distantly semi-supervised setting.

We focus on co-profiling upon every user's *ego network*, which means that we profile a user's attributes and the relationship types of his/her connections given his/her ego network. An ego network is a network surrounding an individual user (a.k.a. the ego), which contains connections between the user and his friends and those among his friends. There are several practical reasons to focus on the ego network of each user instead of a 'full' social network. First, a full social network is unavailable to public in general, whereas a user's ego network is often accessible via certain protocols (e.g., user-authorized applications). Second, an ego network keeps important connections related to one user and typically contains hundreds of nodes, which are much more efficient to deal with than a full one with hundreds of millions of nodes.

As Figure 5.12 illustrates, a user's ego network can be represented as a graph $G = (\mathcal{V}, \mathcal{E})$, where vertices $\mathcal{V}$ represent users and edges $\mathcal{E}$ represent their connections. $\mathcal{V}$ contains the ego user v_0 and his friends (e.g., v_1, v_3). $\mathcal{E}$ contains both the connections between the ego and his friends (e.g., e_{02}, e_{03}) and the connections among their friends (e.g., e_{13}, e_{23}). We specifically use *friends* to refer to all users but the ego user v_0. Let $\mathcal{V}'$ denote the friends, and $\mathcal{E}'$ the connections among the friends.

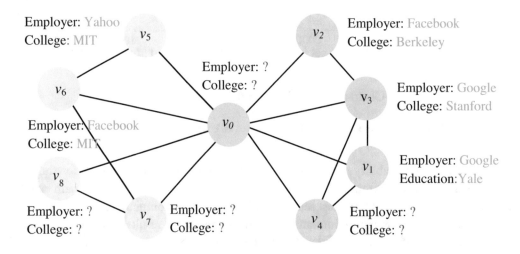

Figure 5.12: An example of an ego social network.

In an ego network, users are linked to relational attributes (e.g., employer, college, location, ages). Let A_k denote the kth attribute (e.g., college), and $a_{i,k}$ the value of the kth attribute of a user v_i (e.g., UIUC). To simplify the discussion, we make two assumptions about an attribute. First, we assume that an attribute (e.g., college) has *categorical* values (e.g., UIUC) from a predefined set (e.g., {Stanford, UIUC, Berkeley}). A numerical attribute (e.g., age) can be categorized (e.g., age-group). Further, we assume that a user has a *single* value (e.g., UIUC) for an attribute (e.g., college). For some attribute (e.g., employer), a user may have multiple values (e.g., v_2 used to work at Google, and now works Facebook now). The solution should be easy to extend for such cases.

Formally, we abstract the problem as follows.

Definition 5.7   User Co-Profiling in Ego Network.   Given a user v_0's ego network $G(\mathcal{V}, \mathcal{E})$, a list of attributes, $A_1, \ldots, A_m$, and some labeled friends $L_k \in \mathcal{V}'$ whose attribute A_k are known, we profile the user v_0's attribute value a_k for each attribute A_k, $k = 1, \ldots, m$, and the latent relationship type between $e_{0x}, v_x \in \mathcal{V}'$.

We note that the labeled friends L_k could be either identical (i.e., a subset of users who provide complete profiles) or different (i.e., different users provide different attributes in their profiles) for different attributes A_k. Both cases are realistic. For the simplicity of the discussion and notation, we assume the former case and denote the labeled friends as L. The solution can handle the second case as well.

5.3.1   OBSERVATIONS

First, we need to properly identify the "correlation" between users' attributes and their social connections, i.e., whether the ego user is likely to share the same value for a given attribute with a friend.

A unique *discrimination correlation* insight is that social connections are "discriminatively" correlated with attributes (e.g., employer, college, interests) through latent relationship types (e.g., colleagues, college classmates, club friends). For example, an ego user (e.g., v_0) is more likely to share the employer (e.g., Google) with his colleagues (e.g., v_3, v_2) than with the college classmates (e.g., v_6, v_5), and he is more likely to share interests (e.g., football) with club friends than his colleagues or classmates. Such an insight can successfully overcome the limitations of the simplistic assumption used by previous methods [Macskassy and Provost, 2003; Pennacchiotti and Popescu, 2011]: that connections are "identically" correlated with attributes (i.e., all the friends are likely to share the same value with the ego for any attribute).

In an ego network, the ego user's connections (e.g., e_{02}) have a one-to-one mapping to his friends (e.g., v_2). We can group friends to *circles* according to the relationship types of their connections. In the rest of the section, we will use a circle to refer to a set of friends who have the same type of connections with the ego.

Since attributes and circles are strongly related (i.e., attributes can help to identify circles, while circles can help to propagate attributes), we jointly profile the ego user's missing attributes and latent circles (relationship types, resp.) of his friends (connections, resp.) given an ego network.

There are two important observations.

- **Attribute-Circle Dependency:** *The friends in a circle share the same value with the ego user for certain attributes.* In a given circle (e.g., colleagues), the friends are much more likely to share the same value for some attribute (e.g., college) with the ego than those in other circles (e.g., college classmates) or in general. Note that friends in a circle (e.g., college classmates) only share one or a few attributes (e.g., college) instead of all attributes.

- **Circle-Connection Dependency:** *Friends across circles are loosely connected.* The average number of connections for each user within the same circle (e.g., ten connections per user within college classmates) is significantly higher than the average number of connections across circles (e.g., less than one connection per user from college classmates to colleagues).

Here we target on identifying a circle (e.g., colleagues) only if the friends in it are likely to share some attributes (e.g., employer). It simplifies the problem. If one intends to classify friends into arbitrary predefined categories (e.g., advisor or advisee), the technique in Section 5.2 could be used.

We should also be aware that some attribute (e.g., name, gender) is associated with no circle or all circles. However, for many important attributes (e.g., employer, college, locations, interests), the above insight is valid. The co-profiling focuses on profiling these circle-related attributes.

5.3.2   MODEL

Intuitively, we can assume that there are K circles in a given ego network, and the friends in each circle share the same value (e.g., UIUC) for certain attributes (e.g., college) with the ego user according to the first dependency and have few connections to the friends in other circles (e.g., colleagues, personal community) according to the second dependency.

K should be a variable to be learned. First, multiple circles may share a common attribute or a circle may share multiple attributes, so K may not be equal to the number of attributes. Second, different ego networks have different numbers of circles. We discuss how to determine K in Section 5.3.3, and let us assume K is fixed for now.

It is also worthwhile to point out that we cannot simply apply standard methods for graph clustering with attributes [McAuley and Leskovec, 2012; Neville et al., 2003; Zhou et al., 2009] for two reasons. First, they assume that node attributes are complete, but attributes of many users are unknown in real data. Second, they assume that nodes in a cluster are similar across all attributes, but friends in a circle only share the same value for one or a few attributes (e.g., college classmates share the same college but have different employers) in our setting.

Now, we introduce how to formally model the above dependencies in an optimization framework. We first introduce the mathematic abstraction of attributes, circles, and their associations.

- To represent attribute values of a user v_i, we introduce an *attribute vector* f_i to represent all the attributes (e.g., employer, college) of v_i. Each dimension of f_i represents a candidate value of certain attribute (e.g., Google for employer or UIUC of college). The candidate values (e.g., Google, Facebook) for an attribute (e.g., employer) can be obtained from the labeled friends or given by a predefined dictionary. The value of the yth dimension of f_i, denoted as $f_{i,y}$, is a real number from 0–1, which indicates how likely v_i is associated with the attribute value represented by the yth dimension. For labeled friends $v_i \in L$, we assign $f_{i,y}$ to 1, if the corresponding attribute value is observed in his profile, and 0, otherwise. For any unlabeled friend $v_i \in U$ or the ego user v_0, f_i is an unknown vector. Further, some circle (e.g., club friends) might be formed based on an unspecified value (e.g., IBM) of a given attribute (e.g., employer) or the values of other attributes (e.g., interests) besides the given attributes (e.g., employer, college). Thus, we add an additional "catchall" dimension in every user's attribute vector and assign a small score (e.g., 0.1) to allow circles to be formed by the catchall dimension. For example, given the college attribute, which has three candidate values (e.g., 1 Berkeley, 2 UIUC, 3 Stanford), and the employer attribute, which has two candidate values (e.g., 4 Google, 5 Facebook), the attribute vector of a user who studies at Berkeley and works for Google is $< 1, 0, 0, 1, 0, 0.1 >$.

- To represent circles of friends, we introduce a latent variable *circle assignment* x_i for each friend $v_i \in V'$. x_i can be any value from 1 to K, which indicates the circle that v_i belongs to. Thus, a circle C_t can be defined as $\{v_i \in V' | x_i = t\}$. We note that, the ego user connects (or belongs) to all the circles in his ego network (e.g., colleagues in Google, college classmates in UIUC), so we do not need the circle assignment for v_0.

- To represent the associations between circles and attributes, we introduce an *association vector* w_t for each circle C_t. The yth dimension of w_t, denoted as $w_{t,y}$, is a binary value, which indicates whether C_t is associated with the attribute value represented by the yth dimension in the attribute vector f_i (i.e., whether the friends in C_t share the corresponding attribute value with the ego). Given the same example mentioned above, the association vector $< 0, 1, 0, 0, 0, 0 >$ of a circle represents that the friends in this circle study in UIUC with the ego user and have different values for other attributes. Since we do not know which attribute values are associated with a circle C_t, w_t is an unknown vector. As friends in a circle only share one or a few attribute values with the ego, w_t should have only one or a few dimensions to be non-zero. To simplify the discussion, let us assume w_i has one dimension equal to one. In most cases, the friends in a circle share one attribute value with the ego (e.g., college classmates only share the same college). We discuss how to extend this assumption in Section 5.3.3.

Now, we model the attribute-circle dependency based on the above abstraction. The dependency suggests that the friends in a circle C_t (in other words, the friends who connect to the ego through the connections of a latent relationship type t) share the attribute value designated by w_t with the ego user v_0. Further, since two connected friends in C_t also have the same relationship type t, they should also share the attribute value designated by w_t. Thus, for every two connected users v_i and v_j in $C_t \cup \{v_0\}$, their attribute vector f_i and f_j should be close on the dimension designated by w_t. Using a standard squared distance function, we tend to minimize $\sum_{v_i \in C_t} (w_t \cdot (f_0 - f_i))^2 + \sum_{e_{ij} \in \mathcal{E}', v_i, v_j \in C_t} (w_t \cdot (f_i - f_j))^2$. Further, the labeled friends L provide explicit knowledge for discovering the associated attribute value of a circle. Specifically, our model should select the associated attribute value for a circle C_t that is shared by many labeled friends in C_t, which is to minimize $\sum_{v_i \in L \cap C_t} (w_t \cdot f_i - 1)^2$.

Then, we model the circle-connection dependency. It suggests that the friends in a circle do not have many connections to the friends in other circles. Thus, our model should minimize the connections across different circles, which is $\sum_{e_{ij} \in \mathcal{E}', x_i != x_j} 1$. This term is the *cut* used by many graph clustering algorithms. Here, we use it to capture of the second dependency. However, as it is only a part of the overall cost function, our model is different from K-cut graph clustering. Note that other graph clustering objectives (e.g., normalized cut or modularity) can also be applied. One reason to choose cut is that circles may not necessarily be balanced.

Finally, we linearly combine the three terms together in the cost function:

$$\lambda_1 \sum_{t=1}^{K} \{ (\sum_{e_{ij} \in E', v_i, v_j \in C_t} (w_t \cdot (f_i - f_j))^2$$

$$+ \sum_{v_i \in C_t} (w_t \cdot (f_0 - f_i))^2) \}$$

$$+ \lambda_2 \sum_{t=1}^{K} \sum_{v_i \in L \cap C_t} (w_t \cdot f_i - 1)^2 + \lambda_3 \sum_{e_{ij} \in E', x_i != x_j} 1. \qquad (5.30)$$

Here, as the weights are free to scale, we confine the weight for the third term λ_3 as one. Next subsection will discuss how to set the parameters λ_1 and λ_2.

5.3.3 INFERENCE ALGORITHM

Now, we derive a co-profiling algorithm. Specifically, we want to design an efficient algorithm, which finds the solutions for unknown variables that jointly minimize Eq. (5.30). There are three groups of unknown variables, circle assignments x_i for each friend $v_i \in \mathcal{V}'$, the attribute vector f_i for each unlabeled friend $v_i \in U$ and the ego user v_0, and the association vector w_t for each circle C_t. Such an objective function is hard to minimize, as it involves not only continuous variables (e.g., f_i) but also discrete variables (e.g., x_i), for which we cannot do derivatives. Based on the idea of the coordinate descent algorithm, we derive the algorithm by iteratively updating each group of unknown variables.

First, we focus on how to update the attribute vector f_i for each unlabeled friend $v_i \in U$ and the ego user v_0, when the circle assignment x_i of each friend $v_i \in V'$ and the association vector w_t for each circle C_t are fixed. The function becomes

$$\sum_{t=1}^{K} \left[\sum_{e_{ij} \in \mathcal{E}', x_i, x_j \in C_t} (w_t \cdot (f_i - f_j))^2 + \sum_{v_i \in C_t} (w_t \cdot (f_i - f_0))^2 \right]$$

which is a quadratic function and can be solved in polynomial time with a standard coordinate descent algorithm. Specifically, we iteratively apply the following update rules until the above function converges. The rules are derived based on finding the first order partial derivative for each unknown variable in the above function, and have the convergence and optimality guarantees.

Since w_t has only one nonzero dimension denoted by y_t, we only use Eq. (5.31) to update f_{i,y_t} of f_i, and do not update other zero dimensions of w_t. Similarly, we use Eq. (5.32) to update f_0 for these non-zero dimensions:

$$f_{i,y_t} = \frac{f_{0,y_t} + \sum_{e_{ij} \in E', v_j \in C_t} f_{j,y_t}}{1 + \sum_{e_{ij} \in E', n_j \in C_t} 1}, v_i \in U \cap C_t, w_{t,y_t} = 1 \tag{5.31}$$

$$f_{0,y_t} = \frac{\sum_{t=1}^{K} \sum_{v_j \in C_t} f_{j,y_t}}{\sum_{t=1}^{K} \sum_{v_j \in C_t} 1}, w_{t,y_t} = 1, \forall t = 1, \ldots K. \tag{5.32}$$

The update functions can be viewed as label propagation from neighbors. While they look similar to graph-based semi-supervised learning [Zhou et al., 2004] or relational classification [Macskassy and Provost, 2003], there are two key differences: (1) the propagation is circle-aware. Attributes are propagated from neighbors only in the same circle rather than all the neighbors; and (2) the propagation is attribute-aware. It only propagates the associated attribute value of the circle instead of all attributes. Thus, the algorithm captures the key insight and is more robust to noises in online social networks.

Second, we discuss how to update the circle assignment x_i for each friend $v_i \in V'$, when the attribute vector f_i for each user $v_i \in V$ and the association vector w_t for each circle C_t are fixed. This can be shown to be an NP-hard problem. We present a heuristic algorithm to find a suboptimal solution for the problem. Intuitively, the algorithm iteratively revises a user v_i's circle assignment x_i to minimize the objective function. Specifically, in every iteration, we use the following equations to decide a friend's circle assignment x_i greedily such that it reduces the value of Eq. (5.30) to the largest degree (or stay unchanged if no other assignment can reduce it):

$$x_i = \arg \max_{t=1,\ldots,K} \left[\sum_{e_{ij} \in E', v_j \in C_t} \left(1 - \lambda_1 (w_t \cdot (f_i - f_j))^2\right) - \lambda_1 (w_t \cdot (f_i - f_0))^2 \right], v_i \in U \tag{5.33}$$

$$x_i = \arg \max_{t=1,\ldots,k} \left[\sum_{e_{ij} \in E', v_j \in C_t} \left(1 - \lambda_1 (w_t \cdot (f_i - f_j))^2\right) \right.$$

$$\left. - \lambda_1 (w_t \cdot (f_i - f_0))^2 - \lambda_2 (w_t \cdot f_i - 1)^2 \right], v_i \in L. \tag{5.34}$$

These update functions can be intuitively explained. Specifically, Eq. (5.33) finds a circle C_t of v_i within which v_i has many connections. The weight of a connection is adjusted according to the difference of the two connected users on the associated attribute value (i.e., $(w_t \cdot (f_i - f_j))^2$). The weight is reduced, if two connected users have different values. λ_1 weights how much the weight of a connection should be reduced according to the difference. Since a connection to a circle should be a positive evidence of belonging to that circle, λ_1 should be less than one. Further, Eq. (5.34) shows that, for a labeled friend v_i, if v_i has the attribute value associated with a circle, v_i should be in this circle. λ_2 weighs the importance of matching on an attribute. As we aim to find attribute-related circles, λ_2 should be large. While the algorithm does not guarantee finding the optimal circle assignments, it improves the solution at every iteration and runs efficiently.

Third, we discuss how to update the association vector w_t for each circle C_t, when each user v_i's attribute vector f_i and the circle assignment x_i are fixed. With the assumption that w_t has only one dimension equal to one, we can exhaustively enumerate all the candidates and select the best w_t. Thus, for each circle, we only need to enumerate a limited number of candidates, which is equal to the number of dimensions of the attribute vector. Further, w_t is independent of each other when x_i and f_i are fixed, so we can decide w_t for each circle independently.

As a whole, the algorithm conducts the above three steps iteratively until it converges. Intuitively, at the first step, it utilizes circles to accurately propagate certain attributes; at the second step, it utilizes the propagated attributes and the network structure to form circles; and at the third step, it automatically finds the associated attribute value for each circle. The algorithm naturally combines attribute prorogation and network clustering with a principled optimization approach.

After the algorithm finishes, for each attribute, we select the candidate attribute value with the maximum value in the ego's attribute vector f_0 as the true value for the attribute. To compare values propagated from different circles and avoid bias to small circles, we compute a final f_0 in the end of iterations without the normalization in Eq. (5.32).

Convergence. Intuitively, at each step, the algorithm tries to improve the cost function and makes sure that the cost function is non-decreasing. Further, as there is a minimum value for the cost function (i.e., 0), the algorithm converges. It might converge to a local optimum, and the results may be affected by different initialization methods. Empirically, this algorithm works well in practice with reasonable or standard initialization. For example, we can initialize each dimension of f_i for an unlabeled friend as 0.5, use a standard graph clustering algorithm to initialize the circle assignments, and then start to update variables iteratively.

Complexity. The complexity of this algorithm is polynomial to the number of the nodes in an ego network, which is very efficient. Specifically, in an iteration, the first step takes $\mathcal{O}(|\mathcal{E}|D)$ to update attribute vectors, where D is the number of dimensions of an attribute vector, the second step takes $\mathcal{O}(|\mathcal{V}|K)$ to update assignments iteratively, and the third step enumerates at most $\mathcal{O}(DK)$ candidates.

Discussion

First, we discuss how to determine the number of circles K. Like many other clustering algorithms, K is an important parameter to learn. We can use a standard technique [Clauset et al., 2004; McAuley and Leskovec, 2012], which tries different values and selects the best one according to an objective function. Since an ego network is small and our algorithm runs efficiently, it is possible for us to test different K. To choose a proper K initially, we can run a standard clustering algorithm first to select an initial value.

Second, we discuss how to extend the algorithm to handle multiple-valued attributes. When we initialize f_i, if one of v_i's attribute values is equal to the value represented by the

y^{th} dimension, we initialize $f_{i,y} = 1$. When we profile the ego user, we can take the top n values from f_0 as its values or use the threshold to determine whether a value should be true or not.

Third, to handle the cases that friends in a circle may share more than one attributes, we could relax the association vector to have multiple nonzero dimensions. A heuristic but fast way is to assign several dimensions of w_t to one together, when each individual dimension achieves similar costs for circle C_t's portion in Eq. (5.30), and then normalize it.

Finally, we explain that the algorithm can handle the attributes that do not induce any circles or induce many circles as good as existing methods (e.g., majority voting). First, for the attributes that do not induce any circles, the algorithm will associate circles to other attributes (if any) or the *catchall* dimension, which we can add to the user attribute vector to generally represent extra attributes. For this case, we can directly run a backup method (i.e., majority voting) to profile users' attribute based on his friends. For the attributes that may induce many circles, the algorithm effectively takes majority voting from the friends in those circles.

5.3.4   EMPIRICAL ANALYSIS

The evaluation dataset contains 175 ego networks from LinkedIn. These ego users are associated with 193 different *colleges*, 375 different *employers*, and 52 different *locations*, and their ego networks contain about 19K users and 110K connections, among which 8K are labeled.

The task is set up to identify two attribute-related circles (relationship types) for each given ego network. They are *colleagues*, in which the friends are likely to share the same employer with the ego user, and *college classmates*, in which the friends are likely to share the same *college* with the ego user. For each ego user in the dataset, a small percentage (e.g., 20%) of his friends are randomly sampled as labeled friends (whose attributes are revealed to the algorithms). Circle assignments (relationship types) for friends (connections) are only used for evaluation.

To measure the performance of a method, we use *precision*, *recall* and *F1*. They are standard metrics used in classification and information retrieval. In this task, precision is the number of friends correctly profiled as belonging to a circle (e.g., colleagues) divided by the total number of friends profiled as belonging to the circle, recall is the number of friends correctly profiled as belonging to a circle divided by the total number of friends in the circle, and F1 is the harmonic mean of precision and recall.

The algorithm CP is compared with three baseline methods.

- RP_a profiles circles in an ego network based on users' attributes. Intuitively, a friend is likely to belong to a target circle (e.g., colleagues), if he/she shares the same value with the ego user for the related attribute (e.g., employer). However, the ego user's attributes are unknown in our setting. Thus, to profile circles based on attributes, RP_a first uses a simple but effective relational classifier [Macskassy and Provost, 2003] to profile the ego's attributes based on his labeled friends. This baseline reflects how well we can profile circles based only on given and profiled attributes.

- RP_n profiles circles in an ego network based on the network structure. Intuitively, friends in a circle (e.g., colleagues) are likely to form a cluster in an ego-network. Thus, we use a modularity based graph clustering method [Clauset et al., 2004] as the second baseline. This method is widely applied in many scenarios (e.g., attribute profiling task [Mislove et al., 2010]) and can automatically select K for a given network. As the algorithm only finds K clusters, RP_n further uses attributes of labeled friends in each cluster to align them with the particular circles in the evaluation set. Intuitively, RP_n selects a cluster, in which the friends are likely to share the related attribute (e.g., employer) of the circle of interest (e.g., colleagues). This baseline shows the performance based only on the network structure.

- RP_{an} is a method that profiles circles in an ego network based on both users' attributes and the network structure [McAuley and Leskovec, 2012]. It assumes that every user has a complete profile, which may fail to handle the challenge that only a few users provide their attributes. The comparison with this baseline helps assess the advantage of CP when dealing with partially available attributes. This method also only outputs K circles, and we apply the strategy used in the previous baseline to identify the two circles (i.e., colleagues and college classmates).

Method	RP_a	RP_n	RP_{an}	CP
Recall	0.13	0.39	0.43	**0.45**
Precision	**0.90**	0.64	0.51	0.69
F1	0.22	0.48	0.46	**0.54**

Figure 5.13: Results for profiling colleagues.

Figure 5.13 shows the results for profiling *colleagues* in terms of precision, recall, and F1. We can obtain the following observations.

- RP_a performs the worst (the lowest F1) among the four methods. Specifically, it has a high precision but a very low recall. The precision is high, indicating that user attributes (e.g., employer) are reliable signals for profiling attribute-related circles (e.g., colleague). The recall is low, because only a few friends have known attributes and RP_a can only make a limited number of predictions. It suggests that, in online social networks, identifying circles with only known attributes is far from enough.

- RP_n performs well in all the three measures. It indicates that the network structure is useful for profiling circles. However, it performs worse than CP in all the three measures, which suggests that the network structure is also insufficient for identifying circles.

- RP_{an} performs similarly as (slightly worse than) RP_n in terms of F1, although it utilizes both user attributes and the network structure. There are two reasons. First, RP_{an} assumes every user has a complete profile, which contains values of different attributes, while values of only one attribute from a few friends are given in our setting. Second, it also does not capture the sparse association between attributes and circles (i.e., one circle is associated with only a few attributes). The results here suggest that available attributes should be carefully used when finding attribute-related circles.

- CP achieves the best F1. It greatly improves RP_a and RP_n because it utilizes both the network structure and users' attributes. Although CP's precision is lower than RP_a, CP's recall is much higher than RP_a. CP also greatly improves RP_{an}, because (1) it is designed to utilize partially available attributes and (2) it models the sparse dependency between attributes and circles.

Method	RP_a	RP_n	RP_{an}	CP
Recall	0.09	0.39	0.43	**0.49**
Precision	**0.98**	0.95	0.94	0.97
F1	0.16	0.55	0.59	**0.65**

Figure 5.14: Results for profiling college classmates.

Figure 5.14 shows the results for profiling *college classmates*. Generally, we can obtain similar findings as the previous experiment. For example, CP significantly improves the three baselines. The results show that the coprofiling approach is general to identify different attribute-related circles.

Further, we investigate the performance of each method using different percentages of friends as labeled friends. Figure 5.15 shows the precision, recall, and F1 for identifying college classmates. We obtain the following observations.

- RP_a: Its recall and F1 increase as the number of the labeled friends increases, since RP_a solely relies on user attributes to profile. Especially, when many friends' attribute values are known, RP_a performs reasonably well. The results again validate that user attributes are strong evidence for identifying attribute-related circles (e.g., colleagues) and should be utilized. However, we should be aware that it is unlikely to have all users' attribute values in many social networks.

- RP_n: It benefits from additional labeled friends slightly, because it mainly utilizes the network structure to group friends (form circles) and only uses friends' attributes to identify the target circle. It performs better than RP_a in most cases, which again confirms the value of the network structure for this task.

- RP_{an}: Its performance also sightly increases as the number of the labeled friends increases, since it also utilizes attributes to identify circles. However, as it assumes that all attributes

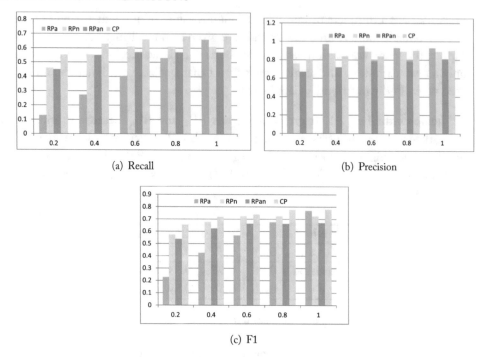

(a) Recall   (b) Precision

(c) F1

Figure 5.15: Results for profiling college classmates with different percentages of labeled friends.

(e.g., employer, college, interests) for every user are known, its performance is not as good as CP.

- CP: Its performance increases as the number of labeled friends increases, which suggests that it can make good use of available attributes. Further, the algorithm performs the best in all the cases, which clearly demonstrates the effectiveness of the coprofiling approach in identifying attribute-related circles.

We can conclude that the CP algorithm can correctly find attribute-related circles (e.g., colleagues). In fact, it can also profile users' attributes based on corresponding circles accurately. Interested readers are suggested to refer to Li et al. [2014] for more details.

CHAPTER 6

Scalable and Robust Topic Discovery

In this chapter, we address the the following computational bottlenecks of topic modeling.

1. **Scalability**. The scale of the problem is determined by these variables: the number of documents D, the vocabulary size V, the total length of documents L, the total number of topics T, and the width of the topical hierarchy K. These variables are not independent. For example, the average length of documents L/D should be larger than 1, and the number of documents D is usually much larger than the vocabulary size V. Typically, the number of tokens L is the dominant factor.

   For scalability the algorithm should have sublinear complexity with respect to L. When the dataset is too large to fit in memory, an ideal algorithm should only perform a small constant number of passes of the data. The traditional inference methods such as Gibbs sampling and EM are expensive, requiring multiple passes of the data. The number of passes has no theoretical bound, and typically needs to be several hundreds or thousands.

2. **Robustness**. In exploration scenarios, part of the constructed hierarchy needs to be revised to customize for users' need. For example, one may want to alter the number of branches or height of a subtree. In Figure 2.1, if one wants to partition topic $t1$ into 3 subtopics instead of 2, but also wants to keep other parts of the tree intact, a robust algorithm should not change the output of topic $t2, t3$ and $t6$ to $t10$.

   Formally, the revision to a subtree $\mathcal{T}(t)$ rooted at topic t is robust, if every topic t' not in the subtree $\mathcal{T}(t)$ remains intact word distribution in the returned hierarchy. This property assures that the local change to a large hierarchy doesn't not alter the remainder of the tree. The revision requires running the construction algorithms for multiple times. The traditional inference The variance of multiple runs can be very large especially when the hierarchy is deep. This prevents a user from robustly revising the local structure of a hierarchy (e.g., changing the number of branches of one node).

We introduce a new hierarchical topic model, which supports divide-and-conquer inference. We then introduce a scalable tensor-based recursive orthogonal decomposition (STROD) method to infer the model. It inherits the nice theoretical properties of the tensor orthogonal decomposition algorithm, but has significantly better scalability. Empirical analysis demonstrates that this method can scale up to datasets that are orders of magnitude larger than the state-of-the-art, while generating quality topic hierarchy that is comprehensible to users.

This chapter follows the same notations in Chapter 2.

6.1 LATENT DIRICHLET ALLOCATION WITH TOPIC TREE

Every document is modeled as a series of multinomial distributions: one multinomial distribution for every non-leaf topic over its child topics, representing the content bias towards the subtopics. For example, in Figure 2.1, there are 4 non-leaf topics: o, $o \odot 1$, $o \odot 2$, and $o \odot 3$. So a document d is associated with 4 multinomial topic distributions: $\theta_{d,o}$ over its 3 children, and $\theta_{d,o\odot1}$, $\theta_{d,o\odot2}$, $\theta_{d,o\odot3}$ over their 2 children each. When the height of the hierarchy $h = 1$, it reduces to the flat LDA model, because only the root is a non-leaf node. Each multinomial distribution $\theta_{d,t}$ is generated from a Dirichlet prior α_t. $\alpha_{t\odot z}$ represents the corpus bias towards z-th child of topic t, and $\alpha_{t\odot0} = \sum_{z=1}^{C_t} \alpha_{t\odot z}$.

For every leaf topic node t, $C_t = 0$, and a multinomial distribution ϕ_t over words is generated from another Dirichlet prior β. These word distributions are shared by the entire corpus.

To generate a word $w_{d,j}$, we first sample a path from the root to a leaf node $o \odot z_{d,j}^1 \odot z_{d,j}^2 \odot \cdots \odot z_{d,j}^h$. The nodes along the path are sampled one by one, starting from the root. Each time one child $z_{d,j}^k$ is selected from all children of $o \odot z_{d,j}^1 \odot \cdots \odot z_{d,j}^{k-1}$, according to the multinomial $\theta_{d,o\odot z_{d,j}^1 \odot \cdots \odot z_{d,j}^{k-1}}$. When a leaf node is reached, the word is generated from the multinomial distribution $\phi_{o\odot z_{d,j}^1 \odot z_{d,j}^2 \odot \cdots \odot z_{d,j}^h}$. Note that the length of the path h is not necessary to be equal for all documents, if not all leaf nodes are on the same level.

The whole generative process is illustrated in Figure 6.1. Table 6.1 collects the notations.

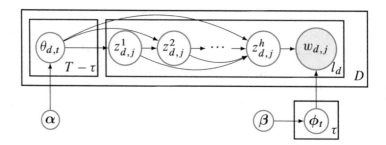

Figure 6.1: Latent Dirichlet allocation with topic tree.

Table 6.1: Notations used in the generative model

Symbol	Description
D	the number of documents in the corpus
V	the number of unique words in the corpus
$w_{d,j}$	the j-th word in the d-th document
l_d	the length (number of tokens) of document d
L	the total number of tokens in the corpus $\sum_{d=1}^{D} l_d$
π_t	the parent topic of topic t
χ_t	the suffix of topic t's notation $(t = \pi_t \odot \chi_t)$
C_t	the number of child topics of topic t
o	the root topic
φ_t	the multinomial distribution over words in topic t
α_t	the Dirichlet hyperparameter vector of topic t
$\theta_{d,t}$	the distribution over child topics of t for document d
T	the total number of topics in the hierarchy
τ	the number of leaf topics in the hierarchy

For every non-leaf topic node, we can derive a word distribution by marginalizing their child topic word distributions:

$$\phi_{t,x} = p(x|t) = \sum_{z=1}^{C_t} p(z|t)p(x|t \odot z) = \sum_{z=1}^{C_t} \frac{\alpha_{t \odot z}}{\alpha_{t \odot 0}} \phi_{t \odot z, x}. \tag{6.1}$$

So in our model, the word distribution ϕ_t for an internal node in the topic hierarchy can be seen as a mixture of its child topic word distributions. The Dirichlet prior α_t determines the mixing weight.

A topical hierarchy $\mathcal{T}$ is parameterized by $\alpha_t(\mathcal{T})$ where $C_t(\mathcal{T}) > 0$, and $\phi_t(\mathcal{T})$ where $C_t(\mathcal{T}) = 0$. We define a topical hierarchy $\mathcal{T}_1$ to be *subsumed* by $\mathcal{T}_2$, if there is a mapping κ from node t in $\mathcal{T}_1$ to node t' in $\mathcal{T}_2$, such that for every node t in $\mathcal{T}_1$, $\pi_t(\mathcal{T}_1) = \pi_{\kappa(t)}(\mathcal{T}_2)$, and one of the following is true:

1. $C_t(\mathcal{T}_1) = C_{\kappa(t)}(\mathcal{T}_2) > 0$ and $\alpha_t(\mathcal{T}_1) = \alpha_{\kappa(t)}(\mathcal{T}_2)$; or

2. $C_t(\mathcal{T}_1) = 0$ and $\phi_t(\mathcal{T}_1) = \phi_{\kappa(t)}(\mathcal{T}_2)$.

In other words, a subsumed tree is obtained by removing all the descendants of some nodes in a larger tree, and absorbing the word distributions of the descendants into the new leaf nodes. Figure 6.2 shows three trees and each tree is subsumed by the one on its right. The subsumed tree retains equivalent high-level topic information of a larger tree, and can be recovered before we recover the larger tree. This idea allows us to recursively construct the whole hierarchy.

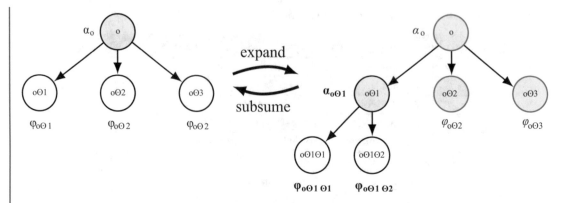

Figure 6.2: An illustration of recursive topical hierarchy construction. The construction order is from left to right. Each time one leaf topic node is expanded into several child topics (unshaded) and the relevant parameters (in bold) are estimated. The same figure explains *subsumption* relationship: a tree on the left is subsumed by a tree on the right.

6.2 THE STROD ALGORITHM

We introduce a Scalable Tensor Recursive Orthogonal Decomposition (STROD) algorithm for topical hierarchy construction, the first that meets the scalability and robustness criteria. It uses tensor (hypermatrix) decomposition to perform moment-based inference of the hierarchical topic model proposed in Section 6.1 recursively.

6.2.1 MOMENT-BASED INFERENCE

The central idea of the inference method is based on the *method of moments*, instead of *maximum likelihood*. It enables tractable computations to estimate the parameters.

In statistics, the *population moments* are expected values of powers of the random variable under consideration. The method of moments derives equations that relate the population moments to the model parameters. Then, it collects empirical population moments from observed samples, and solve the equations using the sample moments in place of the theoretical population moments. In our case, the random variable is the word occurring in each document. The population moments are expected occurrences and co-occurrences of the words. They are related to the model parameters α and ϕ. We can collect empirical population moments from the corpus, and estimate α and ϕ by fitting the empirical moments with theoretical moments. One particular computational advantage is that the inference only relies on the empirical population moments (word co-occurrence statistics). They compress important information from the full data, and require only one scan of the data to collect.

The idea is promising, but not straightforward, to apply to our model. The challenge is that the same population moments can be expressed by parameters on different levels. For the example in Figure 2.1, we can derive equations of the population moments (expected word co-occurrences) based on the model parameters associated with $t1, t2$, and $t3$, or based on those with $t4$–$t9$. Solving these equations independently will find three general topics and six more specific topics, but will neither reveal their relationship, nor guarantee the existence of the relationship.

Below we introduce a recursive inference method step by step.

1. Conditional population moments. We consider the population moments *conditioned* on a non-leaf topic t. The first-order moment is the expectation of word x's occurrence given that it is drawn from topic t's descendant. We have $p(x|t, \alpha) = \sum_{z=1}^{C_t} \frac{\alpha_{t \odot z}}{\alpha_{t \odot 0}} \phi_{t \odot z, x}$ according to Eq. (6.1).

The second-order moment is the expectation of the co-occurrences of two words x_1 and x_2 given that they are both drawn from topic t's descendants:

$$p(x_1, x_2 | t, t, \alpha) = \sum_{z_1 \neq z_2} \frac{\alpha_{t \odot z_1} \alpha_{t \odot z_2}}{\alpha_{t \odot 0}(\alpha_{t \odot 0} + 1)} \phi_{t \odot z_1, x_1} \phi_{t \odot z_2, x_2} + \sum_{z=1}^{C_t} \frac{\alpha_{t \odot z}(\alpha_{t \odot z} + 1)}{\alpha_{t \odot 0}(\alpha_{t \odot 0} + 1)} \phi_{t \odot z, x_1} \phi_{t \odot z, x_2}.$$

(6.2)

Likewise, we can derive the third-order moment as the expectation of co-occurrences of three words x_1, x_2, and x_3 given that they are all drawn from topic t's descendants:

$$
\begin{aligned}
& p(x_1, x_2, x_3 | t, t, t, \alpha) \\
&= \sum_{z_1 \neq z_2 \neq z_3 \neq z_1} \frac{\alpha_{t \odot z_1} \alpha_{t \odot z_2} \alpha_{t \odot z_3}}{\alpha_{t \odot 0}(\alpha_{t \odot 0} + 1)(\alpha_{t \odot 0} + 2)} \phi_{t \odot z_1, x_1} \phi_{t \odot z_2, x_2} \phi_{t \odot z_3, x_3} \\
&+ \sum_{z_1 \neq z_2} \frac{\alpha_{t \odot z_1} \alpha_{t \odot z_2}(\alpha_{t \odot z_1} + 1)}{\alpha_{t \odot 0}(\alpha_{t \odot 0} + 1)(\alpha_{t \odot 0} + 2)} (\phi_{t \odot z_1, x_1} \phi_{t \odot z_1, x_2} \phi_{t \odot z_2, x_3} \\
&+ \phi_{t \odot z_1, x_1} \phi_{t \odot z_1, x_3} \phi_{t \odot z_2, x_2} + \phi_{t \odot z_1, x_3} \phi_{z_1, x_2} \phi_{z_2, x_1}) \\
&+ \sum_{z=1}^{C_t} \frac{\alpha_{t \odot z}(\alpha_{t \odot z} + 1)(\alpha_{t \odot z} + 2)}{\alpha_{t \odot 0}(\alpha_{t \odot 0} + 1)(\alpha_{t \odot 0} + 2)} \phi_{t \odot z, x_1} \phi_{t \odot z, x_2} \phi_{t \odot z, x_3}.
\end{aligned}
$$

(6.3)

These equations exhibit good opportunities for a recursive solution, because the moments conditioned on a topic t can be expressed by only the Dirichlet prior and word distributions associated with its child topics. If these low-order moments can uniquely determine the model parameters, we can use them to recover the child topics of every topic robustly, and by recursion, we can then construct the whole tree (Figure 6.2).

Fortunately, there is indeed a robust technique to recover the parameters from low-order moments.

2. Tensor orthogonal decomposition. Anandkumar et al. [2012] proved that with some mild non-degeneracy conditions, the following equations can be uniquely solved by a tensor orthogonal decomposition method:

$$M_2 = \sum_{z=1}^{k} \lambda_z v_z \otimes v_z, \quad M_3 = \sum_{z=1}^{k} \lambda_z v_z \otimes v_z \otimes v_z,$$

(6.4)

where M_2 is a $V \times V$ tensor (hence, a matrix) and M_3 is a $V \times V \times V$ tensor, λ_z is an unknown positive value about the weight of z-th component v_z, which is an unknown V-dimensional vector. In other words, both M_2 and M_3 can be decomposed into the same number of components, and each component is determined by a single vector. The operator $\otimes$ denotes an outer product between tensors: if $A \in \mathbb{R}^{s_1 \times \cdots \times s_p}$, and $B \in \mathbb{R}^{s_{p+1} \times \cdots \times s_{p+q}}$, then $A \otimes B$ is a tensor in $\mathbb{R}^{s_1 \times \cdots \times s_{p+q}}$, and $[A \otimes B]_{i_1 \ldots i_{p+q}} = A_{i_1 \ldots i_p} B_{i_{p+1} \ldots i_{p+q}}$.

To write Eqs. (6.1)–(6.3) in this form, we define:

$$M_1(t) = \sum_{z=1}^{C_t} \frac{\alpha_{t \odot z}}{\alpha_{t \odot 0}} \phi_{t \odot z} \tag{6.5}$$

$$E_2(t) = [p(x_1, x_2 | t, t, \alpha)]_{V \times V} \tag{6.6}$$

$$M_2(t) = (\alpha_{t \odot 0} + 1) E_2(t) - \alpha_{t \odot 0} M_1(t) \otimes M_1(t) \tag{6.7}$$

$$E_3(t) = [p(x_1, x_2, x_3 | t, t, t, \alpha)]_{V \times V \times V} \tag{6.8}$$

$$U_1(t) = E_2(t) \otimes M_1(t), U_2(t) = \Omega(U_1(t), 1, 3, 2), U_3(t) = \Omega(U_1(t), 2, 3, 1) \tag{6.9}$$

$$M_3(t) = \frac{(\alpha_{t \odot 0} + 1)(\alpha_{t \odot 0} + 2)}{2} E_3(t) + \alpha_{t \odot 0}^2 M_1 \otimes M_1 \otimes M_1$$
$$- \frac{\alpha_{t \odot 0}(\alpha_{t \odot 0} + 1)}{2} [U_1(t) + U_2(t) + U_3(t)], \tag{6.10}$$

where $\Omega(A, a, b, c)$ permutes the modes of tensor A, such that $\Omega(A, a, b, c)_{i_1, i_2, i_3} = A_{i_a, i_b, i_c}$. It follows that:

$$M_2(t) = \sum_{z=1}^{C_t} \frac{\alpha_{t \odot z}}{\alpha_{t \odot 0}} \phi_{t \odot z} \otimes \phi_{t \odot z}, M_3(t) = \sum_{z=1}^{C_t} \frac{\alpha_{t \odot z}}{\alpha_{t \odot 0}} \phi_{t \odot z} \otimes \phi_{t \odot z} \otimes \phi_{t \odot z}.$$

So they fit Eq. (6.4) nicely, and intuitively. If we decompose $M_2(t)$ and $M_3(t)$, the z-th component is determined by the child topic word distribution $\phi_{t \odot z}$, and its weight is $\frac{\alpha_{t \odot z}}{\alpha_{t \odot 0}}$, which is equal to $p(t \odot z | t, \alpha_t)$.

Algorithm 3 outlines the tensor orthogonal decomposition method for recovering the components. It can be partitioned into two parts.

1. Lines 3.1–3.4 project the large tensor $M_3 \in \mathbb{R}^{V \times V \times V}$ into a smaller tensor $\widetilde{T} \in \mathbb{R}^{k \times k \times k}$. $\widetilde{T}$ is not only of smaller size, but can be decomposed into an orthogonal form: $\widetilde{T} = \sum_{z=1}^{k} \widetilde{\lambda}_i \widetilde{v}_i^{\otimes 3}$. $\widetilde{v}_i, i = 1, \ldots, k$ are orthonormal vectors in $\mathbb{R}^k$. This is assured by the whitening matrix W calculated in Line 3.2, which satisfies $W^T M_2 W = I$.

2. Lines 3.5–3.14 perform orthogonal decomposition of $\widetilde{T}$ via a power iteration method. The orthonormal eigenpairs $(\widetilde{\lambda}_z, \widetilde{v}_z)$ are found one by one. To find one such pair, the algorithm randomly starts with a unit-form vector v, runs power iteration (Line 3.9) for n times, and records the candidate eigenpair. This process further repeats by N times, starting from different unit-form vectors, and the candidate eigenpair with the largest eigenvalue is picked

---

Algorithm 3: Tensor Orthogonal Decomposition (TOD)

Input: Tensor $M_2 \in \mathbb{R}^{V \times V}$, $M_3 \in \mathbb{R}^{V \times V \times V}$, number of components k, number of outer and inner iterations N and n

Output: The decomposed components (λ_z, v_z), $z = 1, \ldots, k$

3.1 Compute k orthonormal eigenpairs (σ_z, μ_z) of M_2;

3.2 Compute a whitening matrix $W = M \Sigma^{-\frac{1}{2}}$;

     // $M = [\mu_1, \ldots, \mu_k]$, $\Sigma = diag(\sigma_1, \ldots, \sigma_k)$, $W^T M_2 W = I$

3.3 Compute $(W^T)^+ = M \Sigma^{\frac{1}{2}}$;   // the Moore-Penrose pseudoinverse of W^T

3.4 Compute a k^3 tensor $\widetilde{T} = M_3(W, W, W)$;

     // $\widetilde{T}_{i1,j1,k1} = \sum_{i2,j2,k2} (M_3)_{i2,j2,k2} W_{i2,i1} W_{j2,j1} W_{k2,k1}$

3.5 **for** $z = 1..k$ **do**

3.6 $\lambda^* \leftarrow 0$;   // the largest eigenvalue so far

3.7   **for** *outIter* $= 1..N$ **do**

3.8 $v \leftarrow$ a random unit-form vector;

3.9   **for** *innerIter* $= 1..n$ **do** $v \leftarrow \dfrac{\widetilde{T}(I,v,v)}{\|\widetilde{T}(I,v,v)\|}$;   // power iteration update

3.10   **if** $\widetilde{T}(v, v, v) > \lambda^*$ **then** $(\lambda^*, v^*) \leftarrow (\widetilde{T}(v, v, v), v)$;   // choose the largest eigenvalue

3.11   **end**

3.12 $\lambda_z = \dfrac{1}{(\lambda^*)^2}$, $v_z = \lambda_z (W^T)^+ v^*$;   // recover eigenpair of the original tensor

3.13 $\widetilde{T} \leftarrow \widetilde{T} - \lambda^* v^* \otimes v^* \otimes v^*$;   // deflation

3.14 **end**

3.15 **return** (λ_z, v_z), $z = 1, \ldots, k$

(Line 3.10). After an eigenpair is found, the tensor $\widetilde{T}$ is deflated by the found component (Line 3.13), and the same power iteration is applied to it to find the next eigenpair. After all the k orthonormal eigenpairs $(\widetilde{\lambda}_z, \widetilde{v}_z)$ are found, they can be used to uniquely determine the k target components (λ_z, v_z) (Line 3.12).

The following theorem ensures that the decomposition is unique and fast.

Theorem 6.1   *Assume M_2 and M_3 are defined as in Eq. (6.4), $\lambda_z > 0$, and the vectors v_z's are linearly independent and have unit-form, then Algorithm 3 returns exactly the same set of (λ_z, v_z). Furthermore, the power iteration step of Line 3.9 converges in a quadratic rate.*

Theorem 6.1 relies on several non-trivial claims: (i) the orthogonal decomposition of $\widetilde{T}$ is unique; (ii) the power iteration converges robustly and quickly to the eigenpair; and (iii) a pair of $(\widetilde{\lambda}_z, \widetilde{v}_z)$ uniquely determines a pair of (λ_z, v_z). To see why these claims are non-trivial, we notice that the decomposition of $M_2 = \sum_{z=1}^k \lambda_z v_z \otimes v_z$ is not unique. If (σ_z, μ_z) are orthonormal eigenpairs of M_2, then for any orthonormal matrix $O \in \mathbb{R}^{k \times k}$, $M_2 = \sum_{z=1}^k \sigma_z (O \mu_z) \otimes (O \mu_z)$. So there are infinite number of ways of decomposition if we only consider second-order moments. This explains why CATHY's word co-occurrence network model has no robust inference method, since the word co-occurrence information is equivalent to the second-order moments.

When N and n are sufficiently large, the decomposition error is bounded by the error ϵ of empirical moments from theoretical moments. The number of required inner loop iterations n grows in a logarithm rate with k, and the outer loop N in a polynomial rate. They also proposed possible stopping criterion to reduce the number of trials of the random restart. Since the number of components k is bounded by a small constant $K \approx 10$ in our task, the power iteration update is very efficient, and we observe that $N = n = 30$ are sufficient.

The importance of Theorem 6.1 is that it allows us to use moments only up to the third order to recover the exact components, and the convergence is fast.

3. Recursive decomposition. With Algorithm 3 as a building block, we can divide and conquer the inference of the whole model. We devise Algorithm 4, which recursively infers model parameters in a top-down manner. Taking any topic node t as input, it computes the conditional moments $M_2(t)$ and $M_3(t)$. If t is not root, they are computed from the parent topic's moments and estimated model parameters. For example, according to Bayes's theorem,

$$
\begin{aligned}
[E_2(t)]_{x_1,x_2} &= p(x_1, x_2|t, t, \alpha) \propto p(x_1, x_2, t, t|\pi_t, \pi_t, \alpha) \\
&= p(x_1, x_2|\pi_t, \pi_t, \alpha) p(t, t|x_1, x_2, \pi_t, \pi_t, \alpha) \\
&= [E_2(\pi_t)]_{x_1,x_2} \alpha_{\pi_t \odot \chi_t} (\alpha_{\chi_t \odot z} + 1) \phi_{t,x_1} \phi_{t,x_2} \\
&\Big/ \Bigg(\sum_{z=1}^{C_{\pi_t}} \alpha_{\pi_t \odot z} (\alpha_{\pi_t \odot z} + 1) \phi_{\pi_t \odot z, x_1} \phi_{\pi_t \odot z, x_2} + \sum_{z_1 \neq z_2} \alpha_{\pi_t \odot z_1} \alpha_{\pi_t \odot z_2} \phi_{\pi_t \odot z_1, x_1} \phi_{\pi_t \odot z_2, x_2} \Bigg).
\end{aligned}
$$
$$(6.11)$$

Other quantities in Eqs. (6.5)–(6.10) can be computed similarly. Then it performs tensor decomposition and recovers the parameter α_t and $\phi_{t\odot z}$ for each child topic. It then enumerates its children and makes recursive calls with each of them as input. The recursion stops when reaching leaf nodes, where $C_t = 0$. A call of Algorithm 4 with the root o as input will construct the entire hierarchy.

Algorithm 4: Recursive Tensor Orthogonal Decomposition (RTOD)

Input: topic t, number of outer and inner iterations N, n

4.1 Compute $M_2(t)$ and $M_3(t)$;   // only relies on t's ancestors
4.2 $(\lambda_z, v_z) \leftarrow TOD(M_2(t), M_3(t), C_t, N, n)$;
4.3 $\alpha_{t\odot z} = \alpha_{t\odot 0} \lambda_z, \phi_{t\odot z} = v_z$;
4.4 **for** $z = 1..C_t$ **do**
4.5 $\quad|\quad$ RTOD($t \odot z, N, n$) ;   // Recursion for each subtree
4.6 **end**

The robust revision property is guaranteed due to conditional independence during the recursive construction procedure: (i) once a topic t has been visited in the algorithm, the construction of its children is independent of each other; and (ii) the conditional moments $M_2(t)$ and $M_3(t)$ can be computed independently of t's descendants.

Theorem 6.2   *If $\mathcal{T}_1$ is subsumed by $\mathcal{T}_2$ with the mapping $\kappa(\cdot)$, then the RTOD algorithm on $\mathcal{T}_1$ and $\mathcal{T}_2$ returns identical parameters for $\mathcal{T}_1$ and $\kappa(\mathcal{T}_1)$.*

Therefore, the tree topology can be expanded or varied locally with minimal revision to the inferred topics. This is in particular useful when the structure of the topic tree is not fully determined in the beginning. The recursive construction offers users a chance to see the construction results and interact with the topic tree expansion or its local variations by deciding on the number of topics.

6.2.2   SCALABILITY IMPROVEMENT

Although Algorithms 3 and 4 are robust, they are not scalable. The orthogonal decomposition of the tensor $\widetilde{T} \in \mathbb{R}^{k \times k \times k}$ (Lines 3.5–3.14) is efficient, because k is small. However, the bottleneck of the computation is preparing the tensor $\widetilde{T}$, including Line 4.1 and Lines 3.1–3.4. They involve the creation of a dense tensor $M_3 \in \mathbb{R}^{V \times V \times V}$, and the time-consuming operation $M_3(W, W, W)$. Since V is usually tens of thousands or larger, it is impossible to store such a tensor in memory and perform the tensor product operation. In fact, even the second-order moment $M_2 \in \mathbb{R}^{V \times V}$ is dense and large, challenging both space and time efficiency already.

One solution is to avoid explicit creation of both tensor M_3 and M_2, but explicitly create $\widetilde{T}$ since it is memory efficient. Therefore, the efficient power iteration updates remain as in Algorithm 3. Utilizing the special structure of the tensors in our problem, we show that $\widetilde{T}$ can be created by passing the data only twice, without incurring creations of any dense V^2 or V^3 tensors.

1. **Avoid creating M_2.** For ease of discussion, we omit the conditional topic t in the notation of this and next subsection. According to Eq. (6.7), $M_2 = (\alpha_0 + 1)E_2 - \alpha_0 M_1 \otimes M_1$. E_2 is a sparse symmetric matrix because only two words co-occurring in one document will contribute to the empirical estimation of E_2. However, $M_1 \otimes M_1$ is a full V by V matrix. We would like to compute the whitening matrix W without explicit creation of M_2.

   First, we notice that M_1 is in the *column space* of M_2 (i.e., M_1 is a linear combination of M_2's column vectors), so E_2 has the same column space S as M_2. Also, since $M_2 = \sum_{z=1}^{k} \lambda_z v_z \otimes v_z$ is positive definite, so is $E_2 = \frac{1}{\alpha_0 + 1}(M_2 + \alpha_0 M_1 \otimes M_1)$. Let $E_2 = U\Sigma_1 U^T$ be its spectral decomposition, where $U \in \mathbb{R}^{V \times k}$ is the matrix of k eigenvectors, and $\Sigma_1 \in \mathbb{R}^{k \times k}$ is the diagonal eigenvalue matrix. The k column vectors of U form an orthonormal basis of S. M_1's representation in this basis is $M_1' = U^T M_1$. Now, M_2 can be written as:

$$M_2 = U[(\alpha_0 + 1)\Sigma_1 - \alpha_0 M_1' \otimes M_1']U^T.$$

So, a second spectral decomposition can be performed on $M'_2 = (\alpha_0 + 1)\Sigma_1 - \alpha_0 M'_1 \otimes M'_1$, as $M'_2 = U'\Sigma U'^T$. Then we have:

$$M_2 = UU'\Sigma(UU')^T.$$

Therefore, we effectively obtain the spectral decomposition of $M_2 = M\Sigma M^T$ without creating M_2. Not only the space requirement is reduced (from a dense $V \times V$ matrix to a sparse matrix E_2), but also the time for spectral decomposition. If we perform spectral decomposition for M_2 directly, it requires $\mathcal{O}(V^3)$ time complexity. However, using the twice spectral decomposition trick above, we just need to compute the first largest k eigenpairs for a sparse matrix E_2, and a spectral decomposition for a small matrix $M'_2 \in \mathbb{R}^{k \times k}$. The first decomposition can be done efficiently by a power iteration method or other more advanced algorithms. The time complexity is roughly $\mathcal{O}(k\|E_2\|_0)$, where $\|E_2\|_0$ is the number of non-zero elements in E_2. The second decomposition requires $\mathcal{O}(k^3)$ time, which can be regarded as constant since $k <= K \approx 10$.

2. **Avoid creating M_3.** The idea is to directly compute $\widetilde{T} = M_3(W, W, W)$ without creating M_3. This is possible due to the distributive law: $(A + B)(W, W, W) = A(W, W, W) + B(W, W, W)$. The key is to decouple M_3 as a summation of many different tensors, such that the computation of the product between each tensor and W is easy.

We begin with the empirical estimation of E_3. Suppose we use $c_{i,x}$ to represent the count of word x in document d_i. Then E_3 can be estimated by averaging all the three-word triples in each document:

$$E_3 = \frac{1}{D}[A_1 - A_2 - \Omega(A_2, 2, 1, 3) - \Omega(A_2, 2, 3, 1) + 2A_3]$$

$$A_1 = \sum_{i=1}^{D} \frac{1}{l_i(l_i - 1)(l_i - 2)} c_i \otimes c_i \otimes c_i$$

$$A_2 = \sum_{i=1}^{D} \frac{1}{l_i(l_i - 1)(l_i - 2)} c_i \otimes diag(c_i) \tag{6.12}$$

$$A_3 = \sum_{i=1}^{D} \frac{1}{l_i(l_i - 1)(l_i - 2)} tridiag(c_i),$$

where $tridiag(v)$ is a tensor with vector v on its diagonal: $tridiag(v)_{i,i,i} = v_i$. Let $s_i = \frac{1}{l_i(l_i-1)(l_i-2)}$. From the fact $(v \otimes v \otimes v)(W, W, W) = (W^T v) \otimes (W^T v) \otimes (W^T v) = (W^T v)^{\otimes 3}$, we can derive:

$$A_1(W, W, W) = \sum_{i=1}^{D} s_i(W^T c_i)^{\otimes 3}. \tag{6.13}$$

Based on another fact, $(v \otimes M)(W, W, W) = (W^T v) \otimes M(W, W) = (W^T v) \otimes W^T M W$, we can derive:

$$A_2(W, W, W) = \sum_{i=1}^{D} s_i (W^T c_i) \otimes W^T diag(c_i) W. \tag{6.14}$$

Let W_x^T be the x-th column of W^T, we have:

$$A_3(W, W, W) = \sum_{x=1}^{V} \sum_{i=1}^{D} s_i c_{i,x} (W_x^T)^{\otimes 3}. \tag{6.15}$$

So we do not need to explicitly create E_3 to compute $E_3(W, W, W)$. The time complexity using Eqs. (6.13)–(6.15) is $\mathcal{O}(Lk^2)$, which is equivalent to $\mathcal{O}(L)$ because k is small.

Using the same trick, we can obtain:

$$U_1(W, W, W) = W^T E_2 W \otimes W^T M_1 \tag{6.16}$$
$$(M1 \otimes M1 \otimes M1)(W, W, W) = (W^T M_1)^{\otimes 3}. \tag{6.17}$$

Equation (6.16) requires $\mathcal{O}(k^2 \|E_2\|_0)$ time to compute, while $\|E_2\|_0$ can be large. We can further speed it up.

We notice that $W^T M_2 W = I$ by definition. Substituting M_2 with Eq. (6.7), we have:

$$W^T[(\alpha_0 + 1)E_2 - \alpha_0 M_1 \otimes M_1]W = I \tag{6.18}$$

which is followed by:

$$W^T E_2 W = \frac{1}{(\alpha_0 + 1)}[I + \alpha_0 (W^T M_1)^{\otimes 2}]. \tag{6.19}$$

Pluging Eq. (6.19) into (6.16) further reduces the complexity of computing $U_1(W, W, W)$ to $\mathcal{O}(Vk + k^3)$. $U_2(W, W, W)$ and $U_3(W, W, W)$ can be obtained by permuting $U_1(W, W, W)$'s modes, in $\mathcal{O}(k^3)$ time.

Putting these together, we have the following fast computation of $\widetilde{T} = M_3(W, W, W)$ by passing the data once:

$$\widetilde{T} = M_3(W, W, W) = \frac{(\alpha_0 + 1)(\alpha_0 + 2)}{2} E_3(W, W, W)$$
$$-\frac{\alpha_0(\alpha_0 + 1)}{2}[(U_1 + U_2 + U_3)(W, W, W)] + \alpha_0^2 (W^T M_1)^{\otimes 3} \tag{6.20}$$

which requires $\mathcal{O}(Lk^2 + Vk^2 + k^3)$ time in total.

3. **Estimation of empirical conditional moments.** To estimate the empirical conditional moments for topic t, we compute the 'topical' count of word x in document d_i as:

$$c_{i,x}(t) = c_{i,x} p(t|x) = c_{i,x}(\pi_t) \frac{\alpha_{\pi_t \odot \chi_t} \phi_{t,x}}{\sum_{z=1}^{C_{\pi_t}} \alpha_{\pi_t \odot z} \phi_{\pi_t \odot z, x}} \qquad (6.21)$$

and $c_{i,x}(o) = c_{i,x}$. Then we can estimate M_1 and E_2 using these empirical counts:

$$M_1(t) = \sum_{i=1}^{D} \frac{1}{l_i(t)} c_i(t)$$

$$E_2(t) = \sum_{i=1}^{D} \frac{1}{l_i(t)(l_i(t)-1)} [c_i(t) \otimes c_i(t) - diag(c_i(t))], \qquad (6.22)$$

where $l_i(t) = \sum_{x=1}^{V} c_{i,x}(t)$. These enable fast estimation of empirical moments by passing data once.

Finally, we have a scalable tensor recursive orthogonal decomposition algorithm as outlined in Algorithm 5.

Algorithm 5: Scalable Tensor Recursive Orthogonal Decomposition (STROD)

Input: topic t, number of outer and inner iterations N, n

5.1 Compute $M_1(t)$ and $E_2(t)$ according to Eq. (6.22);

5.2 Find k largest orthonormal eigenpairs (σ_z, μ_z) of E_2;

5.3 $M_1' = U M_1(t)$;   // $U = [\mu_1, \ldots, \mu_k]$, $\Sigma_1 = diag(\sigma_1, \ldots, \sigma_k)$

5.4 Compute spectral decomposition for
$M_2' = (\alpha_{t \odot 0} + 1)\Sigma_1 - \alpha_{t \odot 0} M_1' \otimes M_1' = U' \Sigma U'^T$;

5.5 $M = UU', W = M\Sigma^{-\frac{1}{2}}, (W^T)^+ = M\Sigma^{\frac{1}{2}}$;

5.6 Compute $\widetilde{T} = M_3(W, W, W)$ according to Eq. (6.20);

5.7 Perform power iteration Line 3.5 to 3.14 in Algorithm 3;

5.8 $\alpha_{t \odot z} = \alpha_{t \odot 0} \lambda_z, \phi_{t \odot z} = v_z$;

5.9 **for** $z = 1..C_t$ **do**

5.10 | STROD($t \odot z, N, n$) ;

5.11 **end**

The hyperparameter $\alpha_{t \odot 0}$ and the number of child topics C_t are needed to run the STROD algorithm. Next subsection discusses how to learn them automatically.

6.2.3 HYPERPARAMETER LEARNING

- **Selection of the number of topics.** We discuss how to select C_t when the tree width K is given. We first compute the largest K eigenvalues of E_2 in Line 5.2, and then select the smallest k such that the first k eigenpairs form a subspace that is good approximation of E_2's column space. This is similar to the idea of using Principle Component Analysis (PCA) to select a small subset of the eigenvectors as basis vectors. The *cumulative energy* $g(k)$ for the first k eigenvectors is defined to be $g(k) = \sum_{z=1}^{k} \sigma_z$. And we choose the smallest value of k such that $\frac{g(k)}{g(K)} > \eta$, and let $C_t = k$. $\eta \in [0, 1]$ controls the required energy of the first k eigenvectors, and can be tuned according to the application. When $\eta = 1$ a full K-branch tree will be constructed. When $\eta = 0$ the tree contains a single root node because $C_o = 0$. Typically η between 0.7 and 0.9 results in reasonable children numbers.

- **Learning Dirichlet prior.** First, we note that the individual prior $\alpha_{t,z}$ can be learned by the decomposition algorithm, when the summation $\alpha_{t\odot0}$ of $\alpha_{t\odot1}$ to $\alpha_{t\odot C_t}$ is given to perform the inference for topic t. This already largely reduces the number of hyperparameters that are needed to be given. Large $\alpha_{t\odot0}$ indicates that t's subtopics tend to be mixed together in a document, while small $\alpha_{t\odot0}$ suggests that a document usually talks about only a few of the subtopics. When $\alpha_{t\odot0}$ approaches 0, one expects a document to have only one subtopic of t. So $\alpha_{t\odot0}$ can usually be set empirically according to the prior knowledge of the documents, such as 1–100.

If one wants to learn $\alpha_{t\odot0}$ automatically, here is a heuristic method. Suppose the data are generated by an authentic $\alpha_{t\odot0}^*$, and the moments are computed using the same $\alpha_{t\odot0}^*$, then the decomposition result should satisfy $\sum_{z=1}^{C_t} \alpha_{t\odot z} = \alpha_{t\odot0}$ exactly. However, if one uses a different $\alpha_{t\odot0}$ to compute the moments, the moments could deviate from the true value and result in mismatched $\alpha_{t\odot z}$. The discrepancy between returned $\sum_{z=1}^{C_t} \alpha_{t\odot z}$ and initial $\alpha_{t\odot0}$ indicate how much $\alpha_{t\odot0}$ deviates from the authentic value. So we can use the following fixed-point method to learn $\alpha_{t\odot0}$, where δ is learning rate.

1. Initialize $\alpha_{t\odot0} = 1$.

2. While (not converged):

   (a) perform tensor decomposition for topic t to update $\alpha_{t\odot z}, z = 1, \ldots, C_t$;

   (b) $\alpha'_{t\odot0} = \sum_{z=1}^{C_t} \alpha_{t\odot z}$;

   (c) update $\alpha_{t\odot0} \leftarrow \alpha_{t\odot0} + \delta(\alpha'_{t\odot0} - \alpha_{t\odot0})$.

6.3 EMPIRICAL ANALYSIS

Datasets.

- DBLP title: 1.9M titles, 152K unique words, and 11M tokens.

- CS abstract: A dataset of computer science paper abstracts from Arnetminer.[1] The set has 529K papers, 186K unique words, and 39M tokens.

[1]http://www.arnetminer.org

- TREC AP news: A TREC news dataset (1998). It contains 106K full articles, 170K unique words, and 19M tokens.

- Pubmed abstract: A dataset of life sciences and biomedical topic. 1.5M abstracts[2] from Jan. 2012 to Sep. 2013. The dataset has 98K unique words after stemming and 169M tokens.

We remove English stopwords from all the documents. Documents shorter than three tokens are not used for computing the moments because we rely on up to third order moments.

Methods for comparison.

- hPAM—parametric hierarchical topic model. The hierarchical Pachinko Allocation Model [Mimno et al., 2007] is a state-of-the-art parametric hierarchical topic modeling approach. hPAM outputs a specified number of supertopics and subtopics, as well as the associations between them.

- nCRP—nonparametric hierarchical topic model. Although more recently published non-parametric models have more capability in document modeling, their scalability is worse than nested Chinese Restaurant Process [Griffiths et al., 2004]. So we choose nCRP to represent this category. It outputs a tree with a specified height, but the number of topics is determined by the algorithm. A hyperparameter can be tuned to generate an approximately identical number of topics as other methods.

- splitLDA—recursively applying LDA. A recursive method described by Pujara and Sko-moroch [2012]. It uses LDA to infer topics for each level, and split the corpus according to the inferred results to produce a smaller corpus for inference with the next level. The LDA inference is implemented on top of a fast single-machine LDA inference algorithm [Yao et al., 2009].

- CATHY—recursively clustering word co-occurrence networks. The method we introduced in Section 2.1.

- STROD—and its variations RTOD, $RTOD_2$, $RTOD_3$. Several variations are included for analyzing the scalability improvement techniques: (i) RTOD: recursive tensor orthogonal decomposition without scalability improvement (Algorithm 4); (ii) $RTOD_2$: RTOD plus the technique of avoiding creation of M_2; (iii) $RTOD_3$: RTOD plus the technique of avoiding creation of M_3; and (iv) STROD: Algorithm 5 with the full scale-up technique.

6.3.1   SCALABILITY

The first evaluation assesses the scalability of different algorithms.

Figure 6.3 shows the overall runtime in these datasets. STROD is several orders of magnitude faster than the existing methods. On the largest dataset it reduces the runtime from one

[2]http://www.ncbi.nlm.nih.gov/pubmed

or more days to 18 min. CATHY is the second-best method in short documents such as titles and abstracts because it compresses the documents into word co-occurrence networks. But it is still more than 100 times slower than STROD due to many rounds of EM iterations. splitLDA and hPAM rely on Gibbs sampling, and the former is faster because it recursively performs LDA, and considers fewer dependencies in sampling. nCRP is two orders of magnitude slower due to its nonparametric nature.

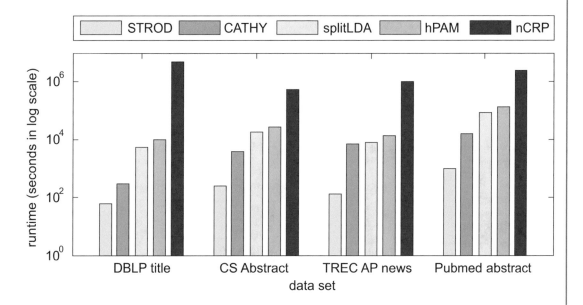

Figure 6.3: Total runtime on each dataset, $h = 2, C_t = 5$.

We now analyze the runtime growth with respect to different factors. Figures 6.4–6.6 show the runtime varying with the number of tokens, tree height, and tree width. We can see that the runtime of STROD grows slowly, and it has the best performance in all occasions. In Figure 6.5, hPAM is excluded because it is designed for $H = 2$. In Figure 6.6, all the methods are assigned with the same number of child topics C_t for each node. nCRP is excluded from all these experiments because it takes too long time to finish (>90 h with 600K tokens).

Figure 6.7 shows the performance in comparison with the slower variations of STROD. Both RTOD and $RTOD_2$ fail to finish when the vocabulary size grows beyond 1K, because the third-order moment tensor M_3 requires $\mathcal{O}(V^3)$ space to create. $RTOD_3$ also has limited scalability because the second order moment tensor $M_2 \in \mathbb{R}^{V \times V}$ is dense. STROD scales up easily by avoiding explicit creation of these tensors.

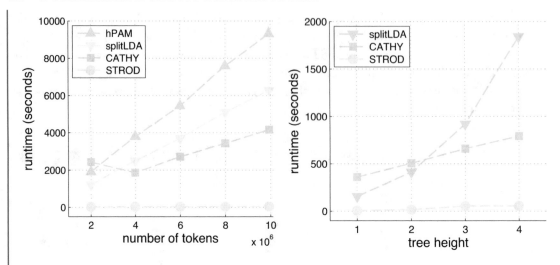

Figure 6.4: Runtime w.r.t corpus size.   **Figure 6.5:** Runtime w.r.t tree height.

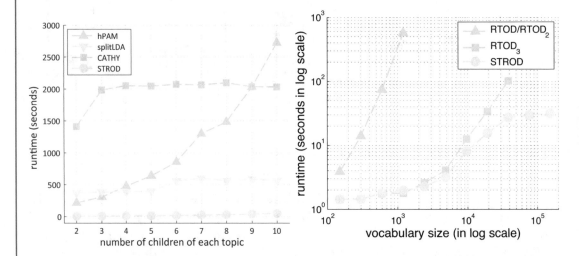

Figure 6.6: Runtime w.r.t the number of children for each topic.

Figure 6.7: STROD scales much better than its variations (all except STROD fail to scale beyond 50K vocabulary size due to memory constraints).

6.3.2   ROBUSTNESS

The second evaluation assesses the robustness of different algorithms. For each dataset, 10,000 documents are sampled and each algorithm is run 10 times and the *variance* among the 10 runs for the same method is measured as follows. Each pair of algorithm runs generate the same number of topics, but their correspondence is unknown. For example, the topic $o \odot 1$ in the first run may be close to $o \odot 3$ in the second run. We measure the KL divergence between all pairs of topics between the two runs, build a bipartite graph using the negative KL divergence as the edge weight, and then use a maximum matching algorithm to determine the best correspondence (top-down recursively). Then we average the KL divergence between matched pairs as the difference between the two algorithm runs. Finally, we average the difference between all $10 \times 9 = 90$ ordered pairs of algorithm runs as the final variance. nCRP is excluded in this section, since even the number of topics is not a constant after each run.

Table 6.2 summarizes the results: STROD has lowest variance. The other three methods based on Gibbs sampling have variance larger than 1 in all datasets, which implies that the topics generated across multiple algorithm runs are considerably different.

We can also look into the variance of STROD without various number of outer and inner iterations N and n. As shown in Figure 6.8, the variance of STROD quickly diminishes when the number of outer and inner iterations grow to 10. This validates the theoretical analysis of their fast convergence and the guarantee of robustness.

In conclusion, STROD achieves robust performance with small runtime. It is stable and reliable to be used as a hierarchy construction method for large text collections.

Table 6.2: The variance of multiple algorithm runs in each dataset

Method	DBLP title	CS abstract	TREC AP news
hPAM	5.578	5.715	5.890
splitLDA	3.393	1.600	1.578
CATHY	17.34	1.956	1.418
STROD	0.6114	0.0001384	0.004522

6.3.3 INTERPRETABILITY

The final evaluation assesses the interpretability of the constructed topical hierarchy, via human judgment. DBLP titles and TREC AP news are used for case study. For simplicity, the number of subtopics are set to be five for all topics. For all the methods, the same phrase mining and ranking procedure are used to enhance the interpretability. nCRP is excluded in this study because hPAM has been shown to have superior performance of it [Mimno et al., 2007].

The topic intrusion task which was introduced in Section 2.3.2 is used in order to evaluate the topic coherence and parent-child relationship.

160 Topic Intrusion questions are randomly generated from the hierarchies constructed by these methods. We consider a higher match between a given hierarchy and human judgment to imply a higher quality hierarchy. For each method, we report the F1 measure of the questions answered 'correctly' (matching the method) and consistently by three human judgers.

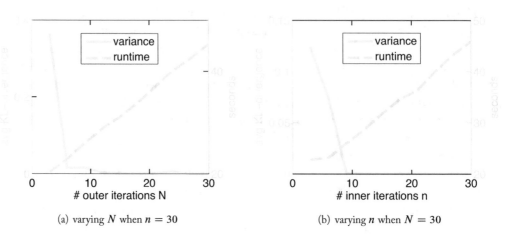

(a) varying N when $n = 30$ (b) varying n when $N = 30$

Figure 6.8: The variance and runtime of STROD when varying # outer and inner iterations N and n (CS abstract).

Figure 6.9 summarizes the results. STROD is among the best performing methods in both tasks. This suggests that the quality of the hierarchy is not compromised by the strong scalability and robustness of STROD.

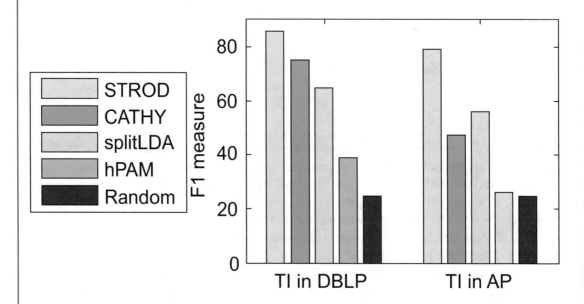

Figure 6.9: Topic intrusion study.

A subset of the hierarchy constructed from CS abstract is presented in Figure 6.10. For each non-root node, it shows the top ranked phrases. Node $o \odot 1$ is about "data," while its children involves database, data mining and bioinformatics. The lower the level is, the more pure the topic is, and the more multigrams emerge ahead of unigrams in general.

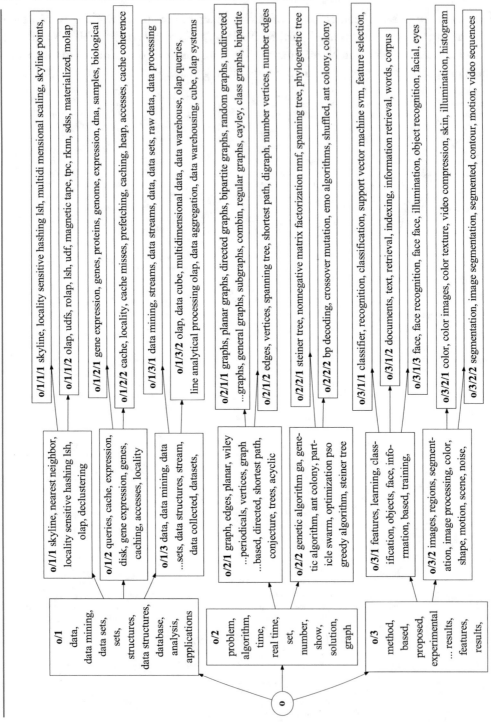

Figure 6.10: Sample of hierarchy generated by STROD.

CHAPTER 7

Application and Research Frontier

This book investigates the problem of latent entity structure mining and introduces methodologies to solve it. It first introduces and presents the background and preliminaries of the latent entity structure mining problems in unstructured and interconnected data, and then categorizes the two important structures, namely the topical and relational structures. Four subtasks are identified and both effective and efficient solutions are proposed and empirically validated on multiple real-world datasets, such as scientific publications, news articles, web pages, and social media.

The solution for the topical structure mining is a breakthrough in terms of quality and computational efficiency. The topics constituted by phrases and entities are much easier to interpret than traditional unigram-based topics. The first theoretically guaranteed method is developed to discover latent topic hierarchy recursively, and solve critical scalability challenges. The algorithm runs with several orders of magnitude faster speed than the state-of-the-art.

The solution for the relational structure mining is presented in a series of studies, and reached leading performance in various domain tasks including academic publications, online forums, and web documents. Both unsupervised and supervised scenarios are covered. Heterogeneous signals including constraints and dependencies are categorized according to semantics, and formulated with simple models.

In the following, we discuss several applications to showcase the impact of the mined structures, and the research frontier.

7.1 APPLICATION

This section describes several representative applications related to mining latent entity structures.

7.1.1 ONLINE ANALYTICAL PROCESSING OF INFORMATION NETWORKS

Using the latent structure mining methods, we can create well-structured information networks from unstructured and loosely structured data. Much knowledge can be derived and explored with such an information network with effective and scalable data-intensive information network analysis technologies.

We take news data as an example. News data is one of the most abundant and well-understood data sources. Unlike research publication datasets (e.g., DBLP) or most relational database data, news data is largely unstructured. Using the techniques discussed in this book, we can (i) mine phrases and construct a topical hierarchy; (ii) discover entity roles and relations within topics of different granularity; and so on.

As a result, we can construct a quality news information network. Such a network can be viewed from two different perspectives: (1) heterogeneous information network with rich text and (2) multi-dimensional hierarchical data cube with rich text. Clearly, modeling a heterogeneous network captures richer node and link semantics and generates better mining result than a homogeneous network or unstructured text collection. Meanwhile, a multidimensional data cube enables us to carry out advanced online analytical processing (OLAP) operations at different levels of granularity. We can explore leveraging both information network and multi-dimensional data cube structures, and conducting network-based mining algorithms at different levels of granularity of the network.

A NewsNetExplorer system is built based on this idea [Tao et al., 2014]. The following outlines a few functional modules with application examples.

Promotion Analysis in News Cube

A cell in the text cube aggregates a set of documents with matching dimension values on a subset of dimensions. Given a keyword query on the news, we want to enable people to find most relevant cells in the data cube. A most relevant cell needs to be generated from two aspects: (i) it is a combination of relevant dimensions; and (ii) it needs to find the best level of granularity of each dimension to describe the cell. For a query like 'Healthcare bill', we may need to find a cell as '{Organization: Congress, Time: 2009}', while for a query like 'Nelson Mandela', a '{Time:Dec/2013, Topic:Death}' cell is better. The top-ranked cells should not only be highly relevant, but also be significant for the query. A relevance scoring model and efficient ranking algorithm has been proposed by Wu et al. [2010b]. It optimizes the search order and prunes the search space by estimating the upper bounds of relevance scores in the corresponding subspaces, so as to explore as few cells as possible for finding top-k answers. An example on the news dataset to generate top-k cells is shown in Figure 7.1.

Entity Ranking on Different Granularity Levels

Entity ranking is an important feature to help readers understand different news lines better without going through the related articles. People are interested in questions like "who are the most influential people in 2013," "what are the most relevant organizations about 911 attack" and "what are the major topics of China during 2000 to 2010." Given a cell of the data cube, the goal

Rank	Cell	#Document	Avg-Relevance
1	Time:Nov-2009, Organization:House of Rep.	93	2.1201770279997136
2	Time:2010, Organization:White House	51	1.9302240668558608
3	Time:2013, Topic:Healthcare.gov rollout	20	1.9351981639862061

Figure 7.1: Top-ranked cells for the query 'Healthcare', the top results correspond to three events: (1) House of Representatives passed the bill, (2) the President signed the bill, and (3) the website rollout.

is to find the top-ranked entities under the cell condition. The role mining technique in Chapter 4 can be applied for entity ranking in different topics.

Example 7.1   Entity ranking in NewsNetExplorer   Given a cell (e.g., 2013 and Iran Nuclear Crisis), perform and show top-k objects of each type (i.e., Person, Topic, Organization, Time, and Location). Drill down to 2013-Nov, Iran Nuclear Crisis or roll up to 2013, International Affairs to see the new ranking result for each type.

7.1.2   SOCIAL INFLUENCE AND VIRAL MARKETING

Word-of-mouth or viral marketing differentiates itself from other marketing strategies because it is based on trust among individuals' close social circle of families, friends, and co-workers. Research shows that people trust the information obtained from their close social circle far more than the information obtained from general advertisement channels such as TV, newspaper, and online advertisements [Nail, 2004]. Thus, many people believe that word-of-mouth marketing is the most effective marketing strategy (e.g., [Misner, 1999]).

The increasing popularity of many online social network sites, such as Facebook, Myspace and Twitter, presents new opportunities for enabling large-scale and prevalent viral marketing online. Consider the following hypothetical scenario as a motivating example. A small company develops an online application and wants to market it through an online social network. It has a limited budget such that it can only select a small number of initial users in the network to use it (by giving them gifts or payments). The company wishes that these initial users would love the application and start influencing their friends on the social network to use it, and their friends would influence their friends' friends and so on, and thus through the word-of-mouth effect a large population in the social network would adopt the application. The problem is whom to select as the initial users so that they eventually influence the largest number of people in the network. The above problem, called *influence maximization*, is first formulated as a discrete optimization problem by Kempe et al. [2003], and has numerous follow up studies.

Influence maximization assumes information cascade as the information propagation model. Information cascade is essentially a tree structure of information propagation. The information receivers in the cascade form a hierarchical relationship. So mining hierarchical relationship can help identify the cascade, and provide accurate input to the influence maximization problem. For example, a social influence analysis method can use the cascade to determine the influence probabilities [Rodriguez et al., 2012].

Adopting the influence maximization idea in word networks, we can summarize the important topical words of a corpus. Further analysis of the influenced words by these topical words reveal the relations of the words. It provides an alternative solution to mining hierarchical relations and hierarchical communities [Wang et al., 2013c].

7.1.3   RELEVANCE TARGETING

Relevance targeting analyzes the data in online social networks and recommends users with ads and other information based on user interests. Relevance is the most important criterion in recommendation. We use ads for the illustration.

The foundation of pay-per-click advertising systems is a system for predicting the click-through rate of an ad for a given user or query. Typically, such predictions are based on a machine learning model that uses various hand-crafted features, and is trained on historical click data. Features employed in current social networking sites mainly inherit the keyword-targeting prototype that is successful in search engine advertising, and demography-targeted advertising that is prevalent in traditional brand advertising. However, the characteristics of social networking sites are not fully exploited by these methods. On one hand, the large amounts of data about user activity and social context are unique in social network services and offer the opportunity for better inference of personalized interests than traditional methods. For example, we can use user-generated content as contextual signals for content-based advertising. On the other hand, it is nontrivial to utilize these data. First, not all the contents have strong contextual relevance. For example, when users check friends' news or talk with friends, their intent is usually not related to business. Second, there is scalability issue if we simply concatenate all the text from linked objects to augment a user's profile. Third, if we use keyword or concept matching to measure the relevance independently of the targeting problem, simply counting the exact matching concepts or even related concepts in the taxonomy cannot capture the latent signal—for example, an ad about photographers is better targeted to users interested in weddings, rather than users interested in photography.

The hierarchical topics are useful to describe the interest of users and ads, so that they can be matched in a common feature space. In other words, mining the roles of users and ads in the hierarchical topics is the foundation of online targeting.

In a heterogeneous social network, users and ads are linked to different types of sources. The information in these sources does not always provide useful knowledge for targeting. It is important to mine the role of different sources in online targeting.

The underlying relations between the concepts in user space and ad space is also useful for relevance matching. For example, a user interest in wedding may have close relation with an ad concept photographer.

In Wang et al. [2011a], a learning method is applied to the social network data in Facebook. It relies on an existing topic hierarchy DMOZ/ODP hierarchy[1] as the feature space. It is expensive to maintain such a hierarchy by human. In the domain where an existing topic hierarchy is not available, obsolete or incomplete, the automatic learning of topic hierarchy will benefit such a targeting application.

7.2   RESEARCH FRONTIER

Many real-world challenges remain unsolved. A long-term research goal is to build a well-structured "information map" for human-friendly navigation, search, and reference. This has potential applications in a wide variety of disciplines, such as software engineering, bioinformatics, and information security, in which data-driven analysis is rising as an important research approach. For example, can we discover the latent structures of genes, proteins and diseases, so that the navigation of these entities on the "information map" could be as easy as location navigation on a geographic map?

A few research directions can be pursued.

1. User-guided structure discovery. The best information structure for different persons may vary according to their background knowledge and preference. For a specific domain like medicine, expert knowledge can be valuable to guide the mining process. Collaboration with specialists in HCI, AI, NLP, and other related fields, may enrich the experience of human-computer cooperation.

2. Evolutionary structures. The information in a certain domain may change at varying speeds, such as stock prices in finance data, or status updates in social media. Accurately analyzing such data requires swiftly adapting to the data stream and constantly incorporating up-to-date information. The infrastructure aspect is critical to building such a swift system, requiring joint effort with experts in systems, networks, and databases.

[1]http://www.dmoz.org

3. Structured information expansion, e.g., via web and crowdsourcing. There are a number of data collections containing structured information, such as knowledge bases like Wikipedia, or closed domain data repositories like Data.gov. It is possible to enrich the structured information provided by these collections with the less structured but high-volume open data from the web or crowdsourcing. Future work would greatly benefit from cross-domain collaboration, since web and crowdsourcing are already common subjects of study for researchers in different fields.

Bibliography

Anima Anandkumar, Rong Ge, Daniel Hsu, Sham M Kakade, and Matus Telgarsky. Tensor decompositions for learning latent variable models. *arXiv preprint arXiv:1210.7559*, 2012. 119

Philip Bille. A survey on tree edit distance and related problems. *Theor. Comput. Sci.*, 337:217–239, June 2005. DOI: 10.1016/j.tcs.2004.12.030. 98

Christopher M. Bishop. *Pattern Recognition and Machine Learning*. Springer, 2006. 82

D. M. Blei and J. D. Lafferty. Visualizing topics with multi-word expressions. *arXiv preprint arXiv:0907.1013*, 2009. 56

Richard H. Byrd, Jorge Nocedal, and Robert B. Schnabel. Representations of quasi-newton matrices and their use in limited memory methods. *Math. Program.*, 63:129–156, 1994. DOI: 10.1007/BF01582063. 97

Jonathan Chang, Jordan Boyd-Graber, Chong Wang, Sean Gerrish, and David M. Blei. Reading tea leaves: How humans interpret topic models. In *NIPS*, 2009. 28, 39

Zheng Chen, Suzanne Tamang, Adam Lee, Xiang Li, Wen-Pin Lin, Javier Artiles, Matthew Snover, Marissa Passantino, and Heng Ji. Cuny-blender tac-kbp2010 entity linking and slot filling system description. In *Proc. 2010 NIST Text Analytics Conference (TAC '10)*, 2010. 100

Kenneth Church, William Gale, Patrick Hanks, and Donald Kindle. Chapter 6. Using statistics in lexical analysis. *Using Statistics in Lexical Analysis*, page 115, 1991. 51

A. Clauset, M.E.J. Newman, and C. Moore. Finding community structure in very large networks. *Physical Review E*, 70(6), 2004. DOI: 10.1103/PhysRevE.70.066111. 110, 112

Bonaventura Coppola, Alessandro Moschitti, and Daniele Pighin. Generalized framework for syntax-based relation mining. In *ICDM*, pages 153–162, 2008. DOI: 10.1109/ICDM.2008.153. 86

Marina Danilevsky, Chi Wang, Nihit Desai, and Jiawei Han. Entity role discovery in hierarchical topical communities. In *KDD Workshop MDS*, 2013.

Marina Danilevsky, Chi Wang, Nihit Desai, Jingyi Guo, and Jiawei Han. Automatic construction and ranking of topical keyphrases on collections of short documents. In *SDM*, 2014. DOI: 10.1137/1.9781611973440.46. 58

Ahmed El-Kishky, Yanglei Song, Chi Wang, Clare R. Vossand, and Jiawei Han. Scalable topical phrase mining from text corpora. *VLDB*, 2015. 58

Brendan J. Frey. *Graphical Models for Machine Learning and Digital Communication*. MIT Press, Cambridge, MA, USA, 1998. 82

T. Griffiths, M. Jordan, J. Tenenbaum, and David M. Blei. Hierarchical topic models and the nested chinese restaurant process. *NIPS*, 2004. 128

Michael A.K. Halliday et al. Lexis as a linguistic level. *In Memory of JR Firth*, pages 148–162, 1966. 41

J.M. Hammersley and P. Clifford. Markov field on finite graphs and lattices, 1971. Unpublished.

Kalervo Järvelin and Jaana Kekäläinen. Cumulated gain-based evaluation of ir techniques. *ACM Trans. Inf. Syst.*, 20(4):422–446, October 2002. DOI: 10.1145/582415.582418. 53

Heng Ji and Ralph Grishman. Knowledge base population: Successful approaches and challenges. In *ACL*, 2011. 99

David Kempe, Jon M. Kleinberg, and Éva Tardos. Maximizing the spread of influence through a social network. In *KDD'03 Proceedings of the Ninth ACM SIGKDD International Conference of Knowledge Discovery and Data Mining*. pages 137–146. DOI: 10.1145/956750.956769. 138

Frank R. Kschischang, Senior Member, Brendan J. Frey, and Hans andrea Loeliger. Factor graphs and the sum-product algorithm. *IEEE Trans. Inform. Theor.*, 47:498–519, 2001. DOI: 10.1109/18.910572. 82

Qi Li, Heng Ji, and Liang Huang. Joint event extraction via structured prediction with global features. In *ACL*, 2013. 29

Rui Li, Shengjie Wang, Hongbo Deng, Rui Wang, and Kevin Chen-Chuan Chang. Towards social user profiling: unified and discriminative influence model for inferring home locations. In *KDD*, pages 1023–1031, 2012. DOI: 10.1145/2339530.2339692. 103

Rui Li, Chi Wang, and Kevin Chen-Chuan Chang. User profiling in an ego network: co-profiling attributes and relationships. In *WWW*, pages 819–830, 2014. DOI: 10.1145/2566486.2568045. 114

Robert V. Lindsey, William P. Headden, III, and Michael J. Stipicevic. A phrase-discovering topic model using hierarchical pitman-yor processes. In *EMNLP-CoNLL*, 2012. 56

Sofus A. Macskassy and Foster Provost. A simple relational classifier. In *Proceedings of the Second Workshop on Multi-Relational Data Mining (MRDM-2003) at KDD-2003*, pages 64–76, 2003. 105, 109, 111

Julian McAuley and Jure Leskovec. Learning to discover social circles in ego networks. In *NIPS*, pages 548–556. 2012. DOI: 10.1145/2556612. 106, 110, 112

David Mimno, Wei Li, and Andrew McCallum. Mixtures of hierarchical topics with pachinko allocation. In *ICML*, 2007. DOI: 10.1145/1273496.1273576. 128, 131

Alan Mislove, Bimal Viswanath, Krishna P. Gummadi, and Peter Druschel. You are who you know: inferring user profiles in online social networks. In *WSDM '10*, pages 251–260, 2010. DOI: 10.1145/1718487.1718519. 103, 112

Ivan R. Misner. *The World's Best Known Marketing Secret: Building Your Business with Word-of-Mouth Marketing*. 2nd ed., 1999, Bard Press. 137

Jim Nail. The consumer advertising backlash, May 2004. Forrester Research and Intelliseek Market Research Report. 137

Jennifer Neville, Micah Adler, and David Jensen. Clustering relational data using attribute and link information. In *Proceedings of the Text Mining and Link Analysis Workshop*, 2003. 106

David Newman, Jey Han Lau, Karl Grieser, and Timothy Baldwin. Automatic evaluation of topic coherence. In *NAACL-HLT*, 2010. 28, 29

Ted Pedersen. Fishing for exactness. *arXiv preprint cmp-lg/9608010*, 1996. 51

Marco Pennacchiotti and Ana-Maria Popescu. Democrats, republicans and starbucks afficionados: user classification in twitter. In *KDD*, pages 430–438, 2011. DOI: 10.1145/2020408.2020477. 105

Jay Pujara and Peter Skomoroch. Large-scale hierarchical topic models. In *NIPS Workshop on Big Learning*, 2012. 128

Manuel G. Rodriguez, Jure Leskovec, and Andreas Krause. Inferring networks of diffusion and influence, *ACM Transactions on Knowledge Discover from Data*, 5(4), Art. No. 21, 2012. 138

Gideon Schwarz. Estimating the dimension of a model. *Ann. Statist.*, 6(2):461–464, 1978. DOI: 10.1214/aos/1176344136. 10

J. Seo, W.B. Croft, and D.A. Smith. Online community search using thread structure. In *CIKM '09*, 2009. DOI: 10.1145/1645953.1646262. 101

Padhraic Smyth. Model selection for probabilistic clustering using cross-validated likelihood. *Statistics and Computing*, 10(1):63–72, 2000. DOI: 10.1023/A:1008940618127. 27

Yizhou Sun, Yintao Yu, and Jiawei Han. Ranking-based clustering of heterogeneous information networks with star network schema. In *KDD*, 2009. DOI: 10.1145/1557019.1557107. 13, 34

Jie Tang, Jimeng Sun, Chi Wang, and Zi Yang. Social influence analysis in large-scale networks. In *KDD*, 2009. DOI: 10.1145/1557019.1557108.

Fangbo Tao, George Brova, Jiawei Han, Heng Ji, Chi Wang, Brandon Norick, Ahmed El-Kishky, Jialu Liu, Xiang Ren, and Yizhou Sun. Newsnetexplorer: Automatic construction and exploration of news information networks. In *SIGMOD*, 2014. DOI: 10.1145/2588555.2594537. 136

Takashi Tomokiyo and Matthew Hurst. A language model approach to keyphrase extraction. In *Proceedings of the ACL 2003 Workshop on Multiword Expressions: Analysis, Acquisition and Treatment - Volume 18*, MWE '03, 2003. DOI: 10.3115/1119282.1119287. 40, 41

Chi Wang, Jiawei Han, Yuntao Jia, Jie Tang, Duo Zhang, Yintao Yu, and Jingyi Guo. Mining advisor-advisee relationships from research publication networks. In *KDD*, 2010. DOI: 10.1145/1835804.1835833.

Chi Wang, Rajat Raina, David Fong, Ding Zhou, Jiawei Han, and Greg Badros. Learning relevance from heterogeneous social network and its application in online targeting. In *SIGIR*, 2011a. DOI: 10.1145/2009916.2010004. 139

Chi Wang, Jie Tang, Jimeng Sun, and Jiawei Han. Dynamic social influence analysis through time-dependent factor graphs. In *ASONAM*, 2011b. DOI: 10.1109/ASONAM.2011.116.

Chi Wang, Wei Chen, and Yajun Wang. Scalable influence maximization for independent cascade model in large-scale social networks. *Data Mining Knowledge Discov.*, 25(3):545–576, 2012. DOI: 10.1007/s10618-012-0262-1.

Chi Wang, Marina Danilevsky, Nihit Desai, Yinan Zhang, Phuong Nguyen, Thrivikrama Taula, and Jiawei Han. A phrase mining framework for recursive construction of a topical hierarchy. In *KDD*, 2013a. DOI: 10.1145/2487575.2487631.

Chi Wang, Marina Danilevsky, Jialu Liu, Nihit Desai, Heng Ji, and Jiawei Han. Constructing topical hierarchies in heterogeneous information networks. In *ICDM*, 2013b. DOI: 10.1007/s10115-014-0777-4.

Chi Wang, Xiao Yu, Yanen Li, Chengxiang Zhai, and Jiawei Han. Content coverage maximization on word networks for hierarchical topic summarization. In *CIKM*, 2013c. DOI: 10.1145/2505515.2505585. 138

Xuerui Wang, Andrew McCallum, and Xing Wei. Topical n-grams: Phrase and topic discovery, with an application to information retrieval. In *ICDM*, 2007. DOI: 10.1109/ICDM.2007.86. 56

Y. Weiss and W. T. Freeman. On the optimality of solutions of the max-product belief propagation algorithm in arbitrary graphs. *IEEE Trans. Inform. Theory*, 47:723–735, 2001. DOI: 10.1109/18.910585. 93

Tianyi Wu, Yuguo Chen, and Jiawei Han. Re-examination of interestingness measures in pattern mining: A unified framework. *Data Min. Knowl. Discov.*, 2010a. DOI: 10.1007/s10618-009-0161-2. 78

Tianyi Wu, Yizhou Sun, Cuiping Li, and Jiawei Han. Region-based online promotion analysis. In *EDBT*, pages 63–74, 2010b. DOI: 10.1145/1739041.1739052. 136

Zi Yang, Jie Tang, Bo Wang, Jingyi Guo, Juanzi Li, and Songcan Chen. Expert2bole: From expert finding to bole search (demo paper). In *KDD'09*, 2009. 86

Limin Yao, David Mimno, and Andrew McCallum. Efficient methods for topic model inference on streaming document collections. In *KDD*, 2009. DOI: 10.1145/1557019.1557121. 128

Xiaoxin Yin and Sarthak Shah. Building taxonomy of web search intents for name entity queries. In *WWW '10*, 2010. DOI: 10.1145/1772690.1772792.

Wayne Xin Zhao, Jing Jiang, Jing He, Yang Song, Palakorn Achananuparp, Ee-Peng Lim, and Xiaoming Li. Topical keyphrase extraction from twitter. In *ACL-HLT*, 2011. 40, 41, 53

Dengyong Zhou, Olivier Bousquet, Thomas Navin Lal, Jason Weston, and Bernhard Sch?lkopf. Learning with local and global consistency. In *NIPS*, pages 321–328, 2004. 109

Yang Zhou, Hong Cheng, and Jeffrey Xu Yu. Graph clustering based on structural/attribute similarities. *Proc. VLDB Endow.* 2(1):718–729, August 2009. DOI: 10.14778/1687627.1687709. 106

Authors' Biographies

CHI WANG

Chi Wang is a researcher at Microsoft Research, Redmond, Washington. He received his Ph.D. degree in computer science from the University of Illinois at Urbana-Champaign in 2014. He graduated from Tsinghua University, China, in 2009. His research has been focused on data mining, information network analysis, and text mining. He is the first winner of the prestigious Microsoft Research Graduate Research Fellowship in the history of Computer Science, University of Illinois at Urbana-Champaign.

JIAWEI HAN

Jiawei Han is the Abel Bliss Professor in the Department of Computer Science at the University of Illinois. His research interests include data mining, information network analysis, and database systems, and he has over 600 publications. He served as the founding Editor-in-Chief of *ACM Transactions on Knowledge Discovery from Data* (TKDD). Jiawei has received the ACM SIGKDD Innovation Award (2004), IEEE Computer Society Technical Achievement Award (2005), IEEE Computer Society W. Wallace McDowell Award (2009), and Daniel C. Drucker Eminent Faculty Award at UIUC (2011). He is a Fellow of ACM and a Fellow of IEEE. He is currently the Director of Information Network Academic Research Center (INARC) supported by the Network Science-Collaborative Technology Alliance (NS-CTA) program of U.S. Army Research Lab. His co-authored textbook *Data Mining: Concepts and Techniques* (Morgan Kaufmann) has been adopted worldwide.

d in the United States
r & Taylor Publisher Services